The Taste for Ethics

The International Library of Environmental, Agricultural and Food Ethics

VOLUME 7

THE TASTE FOR ETHICS

An Ethic of Food Consumption

By

Christian Coff

Centre for Ethics and Law, Copenhagen, Denmark

Translator

Edward Broadbridge

Springer

A C.I.P. Catalogue record for this book is available from the Library of Congress.

ISBN-10 1-4020-4553-0 (HB)
ISBN-13 978-1-4020-4553-0 (HB)
ISBN-10 1-4020-4554-9 (e-book)
ISBN-13 978-1-4020-4554-7 (e-book)

Published by Springer,
P.O. Box 17, 3300 AA Dordrecht, The Netherlands.

www.springer.com

Printed on acid-free paper

Printed in the Netherlands.

CONTENTS

ACKNOWLEDGEMENTS

This book has been written as part of several research projects on food ethics that I have carried out at the Centre for Ethics and Law in Copenhagen, Denmark. I am very grateful to the staff at the centre for the inspiring and stimulating discussions on philosophy, ethics and food – food, however, being mostly discussed during the lunch breaks. I also wish to thank them for pointing to relevant philosophical literature and discussions as well as for their advice on my use of philosophical literature in the context of food and consumption. Without their stimulation, criticism, patience and comments on my work, this book is unlikely to have seen the light of day. I also extend my thanks to Niels Mattsson Johansen, Peter Kemp, Jacob Rendtorff, Lisbeth Withøfft Nielsen, Emily Hartz and not least to Thomas Achen at Linköping University, Sweden, for introducing me to the philosophy of recognition.

The empirical studies mentioned in Chapter Six were conducted in Denmark in 2003 and 2004 in collaboration with Lise Walbom from the Danish Consumers Co-operative Society (FDB). I am grateful to her and the organization for making these rather unconventional studies possible.

Translating from Danish to English, and sometimes also from German and French, is indeed a difficult task and at times seems almost impossible. I am therefore grateful for Edward Broadbridge's translation and proofreading of the book.

Most of this book was written with the financial support from two grants by the Danish Research Council from 2001 to 2004. A small part was written during my time as a coodinator of an EU-supported project entitled *Ethical Traceability and Informed Choice in Food Ethical Issues* under the sixth framework programme, Science and Society.

FOREWORD

This book marks a new departure in ethics. In our culture ethics has first and foremost been a question of "the good life" in relation to other people. Central to this ethic was friendship, inspired by Greek thought (not least Aristotle), and the caritas concept from the Judaeo-Christian tradition. Later moral philosophers also included man's relation to animals, and it was agreed that the mistreatment of animals was morally reprehensible. But no early moral teaching discussed man's relation to the origin of foodstuffs and the system that produced them; doubtless the question was of little interest since the production path was so short. The interest in good-quality food is of course an ancient one, and healthy eating habits have often been underlined as a condition for the good life. But before industrialization the production of this food was easy to follow. As a rule, that is no longer the case. The field of ethics must therefore be extended to cover responsibility for the production and choice of foodstuffs, and it is this food ethic that Christian Coff sets out to trace.

In doing so he shows how the focus of ethics can be expanded from its concern for the good life on earth with and for others to cover the good life in fair food production practices, and how not least through using our integrity or life coherence we can reflect ethically, or caringly, about living organisms, ecological systems and our human identity. Ethics here is not reduced to a merely personal ethic but embraces a nature ethic, an ethic for our physical lives within the whole of nature. And as an "ethic of taste" it deals with our relation to all that we eat – normally not at the moment of eating, and certainly not when we are gathered for a celebratory meal, but when we are purchasing foodstuffs or producing the raw materials ourselves. In practice this means that whoever is involved in the production of food, as a professional producer or merely as a private citizen growing vegetables or keeping chickens, and in particular as the consumer shopper, ought to be on the lookout for food with the healthiest production history behind it. Food ethics is related to agriculture, its production process and its marketing and distribution – and our choices of what to eat. As such, food ethics has very much to do with the safeguarding and promotion of good health, and in this sense it is one of the conditions for a good life.

In Coff's study ethics are extended from personal relationships to the traceability of foodstuffs, and thus to our whole relation to nature as the environment of our lives. His pioneering work presents a new way of thinking when we wish to act responsibly for a healthy and good life. It tells us how as consumers we should consider choosing our foodstuffs, as parents for children, as a cook for institutions and so on. But it also concerns everyone involved in producing and presenting foodstuffs in a modern technological society. *The Taste for Ethics* illuminates a central aspect of the difficulty of being responsible today in the face of a complex production machinery, and in so doing it helps us to become precisely that – responsible.

Dr. Peter Kemp
Professor of Philosophy at the Danish University of Education

PREFACE

Over the last decade or so large numbers of consumers have acquired a taste for ethics. It is those consumers and their possibilities for action that are in focus in this book. And as always with a new subject, there is inevitably a search for definitions and vocabulary: What is food ethics? Is it indeed possible to have ethics for food? And why has the taste for ethics not emerged among consumers before now?

Two factors seem particularly important. First, the abundance of food and the astonishing variety now available to Western consumers make it possible to focus on other questions than the basics of human hunger and the supply of food. Second, the powerful technological development within the life sciences and the risks thereby incurred have brought about new forms of intervention in living nature that have in turn given rise to ethical reflections on food production practices, most notably on the use of gene technology. Especially in Europe this has led to serious controversies between advocates and opponents. Among the latter are many consumers who find it difficult to understand the need for new risky technologies with unforeseen consequences at a time when there is no hunger in the Western world. The advantages for consumers seem small or even totally absent. Another example is the BSE crisis, which has rendered many consumers sceptical or critical about tampering with nature. All in all, food production practices – from farming to food processing – are thus faced not only with various problems linked to the social aspects of farming and food culture but also with a growing concern for the environment and for animal welfare. And it seems unlikely that future technological development will be able to solve so many problems in the short run. By way of immediate response, an ethical reflection is essential for guidance between what is "too much" and "too little" in food production practices. Today, when we are in search of food ethics, we are asking for *the vision of the good life with and for others in fair food production practices*.[1]

[1] This is based on Paul Ricœur's definition of ethics: "Aiming at the good life for and with others in just institutions" (*Oneself as Another*, ch. 7). I have made two changes. First, Ricœur borrows the phrase "aiming at" from Aristotle's definition of ethics (aiming at the good life) in *The Nicomachean Ethics*. I prefer to use "vision of" instead of "aiming at". Second, because the subject here is food ethics, "institutions" has been replaced by "food production practices".

Indeed, it is not an exaggeration to speak of a food *crisis*, not in the sense of a supply crisis but of an ethical crisis. Food researchers and food industries are ready to use the new opportunities stemming from technological development, whereas critical consumers seem to want to slow, or even halt, this development. The designation *crisis* signifies the dissolution of a given order and indicates a formless intermediate position, a turning point or a transformation, before a new order has been established. A crisis creates a situation dominated by instability, the outcome of which by definition is impossible to foresee. The word *crisis* comes from ancient Greek and is derived from the verb to *distinguish* or to *decide*. In Chinese the concept of crisis contains a twofold meaning and therefore consists of two signs: one for *beginning* and one for *end*. If the changes brought about by the crisis are dramatic, they might result in a revolution that is irreversible and in a violent change of the existing order. Food production practices are therefore in crisis, for there is a dissolution of the existing order, characterized by numerous attempts to distinguish between good and bad food production practices and good and bad technological developments.

Food is a subject in the life sciences, agriculture, economy, the culinary art and aesthetics. It is also a subject within sociology, anthropology and psychology, but it has so far only been a very peripheral subject in philosophy – the reason perhaps being that food is a somewhat earth-bound and materialistic subject, difficult to raise to higher philosophical levels than the purely utilitarian considerations of costs and benefits. Utilitarianism is an important aspect of food ethics that it would be foolish to overlook and one that, in the reflections of most people, is an integrated part of any food ethic. But this is certainly not the only way we think about, and relate to, food. Dealing with food ethics in a non-utilitarian way is a venture, and an even bigger venture for an agronomist like myself. *Agro* means soil and it follows that agronomists are occupied with the rules and laws of the soil. I am therefore running the risk of being considered a "peasant philosopher" – the name assigned by the French philosopher Emmanuel Lévinas (1905–1995) to the German philosopher Martin Heidegger (1889–1976). According to Lévinas, Heidegger had no understanding of the relationship between human beings, but only of that between man and nature (nature was called *die Erde, the earth*, by Heidegger).

Grasping one's *own* existence and the understanding of the *self* was the main task of philosophy, said Heidegger. In his view the presence of others does not complement but rather disturbs the existence of the self. Food ethics is of course a relationship between man and nature, but it should not be limited to that and therefore we do not need to follow in the footsteps of Heidegger. I hope it will become clear to readers of this book that food ethics is also a question of the relationship between human beings.

If we dwell briefly on Lévinas and follow his ideas about "peasant philosophy", it is reasonable to state that agriculture is an occupation for those who have no understanding of the relationship between human beings. Looking at the present agricultural practices and their often very poor public image I am tempted to say that Lévinas was right. Today machines have replaced manpower, so that on most farms only a single person tills the soil. In many cases farming has become an isolated occupation with limited social contact. The word "peasant" is used in a condescending manner to refer to somebody stupid or foolish. How has this come about? I believe it is because those who are bright and wise enough have left agriculture behind and devoted themselves to tasks that they consider a bigger challenge to mental activity; they have left the hard, manual grind of tilling the soil to those who *cannot* think. If the idea of the "peasant" in its negative sense is taken seriously, farming and food production is an occupation for those who do not know how to think. Leaving farming and food production to those who cannot think is the same as not giving any thought to food and not paying any attention to it.

This is of course a surprising claim that ought to be amplified. For in a manner of speaking it seems that we are thinking more about food than ever before. We think about food in at least three ways. First, it is seen as part of a social context. The consumption of food usually takes place under social circumstances and contributes to a person's identity or self-understanding and social position. Second, food has an aesthetic dimension as "prepared taste". Food is "prepared aesthetics" when cooked, and "natural" or "non-prepared aesthetics" in its more raw or natural form. A lot of attention is paid to transforming food from its natural to its prepared form. Third, intellectual activity is used to rationalize food production and food processing by scientific and economic means. So one has to ask, in what sense do we not pay any attention to food? The claim should be understood to mean that we do not think about food in its broad context but reduce it to one or more of the areas mentioned. Knowledge about food is often very specific and detailed. For instance, if we consider the huge number of very popular cookbooks available, it is evident that a large amount of detailed knowledge about cooking exists. This is in itself not a problem. It only becomes a problem when it prevents us from a wider understanding of food and food production in societal and environmental contexts – when the detailed knowledge excludes food ethics as the vision of the good life with and for others in just, fair, etc. food production practices.

In Part II of the book I describe how and why these detailed and simultaneously reductionist views of food have become dominant in our culture. I also criticize the narrow understanding of food because in my view this often prevents us from seeing the ethical aspects of food production and consumption. In a sense the book is an attempt to rehabilitate the concept of "peasants", to focus on the positive instead of the negative understanding of the word and to investigate the link between food production practices and philosophy. This means that

I will try to bring together the often rather abstract thoughts of philosophers regarding present food production practices and vice versa, to consider food not only as substance and economics but to give these a history and an ethic.

Foodstuffs have an origin and a history before they are consumed. This history can be known or unknown to the consumer; it can be of importance or of no importance. However, if ethics is to have any meaning in relation to food, it is in the *production history* of the food being processed that it must be found. Food ethics as it is developed here is based on humanist traditions such as phenomenology, hermeneutics and semiology, which together are used in the analysis of our sensuous and reflective relationship to food. This description and analysis leads to the central aim of the book: an investigation of the possibilities for consumers to bring their own food ethics into action. For this reason food ethics is here also developed as a narrative ethics; it is the recounting or the self-experience of the production history that forms the basis for the ethical attitude to food.

Today most people in Western Europe are living in urban areas, and their contact with nature, not to mention agriculture, is therefore rather limited. Living in cities where most things are made for specific purposes, we are tempted to adopt the same kind of thinking in our reflections on nature. We see it in terms of utility and benefit. There is one way, though, in which even city dwellers are in contact with nature every day: through food. Food *comes* from nature; it is *made* from nature. This immediately suggests that food could function as a starting point for a consumer ethics for the natural environment. It is my intention to see how far we can go in thinking of food ethics as a way of mediating an ethic for the environment and nature.

Food is made from nature, but in the preparation and cooking of food nature is transformed into culture. As culture, food is a part of the relations between human beings and as such it is also an intermediary in those relations. But just as much as we are likely to forget that food is made from nature, so are the consequences of food consumption for other people – e.g. those involved in food production – somewhat obscure.

During the writing of this book I was also involved from 1998 to 2002 in the establishment of a consumer-supported agricultural guild called Landbrugslauget, housed close to Copenhagen. This is organized as a shareholder farm: 500 consumers, mainly from Copenhagen, own the farm "Brinkholm" together with the farmers. This makes it possible for the consumers to acquire a deeper understanding of agriculture and food production practices. The consumers get to learn about the production history in another way than they would otherwise have done. Working with this project has given me much inspiration and can in some sense be considered a part of the empirical basis of the present book.

I would like to add a few remarks on the methods I have employed, or more precisely the methods that developed when working with the issue of food ethics. The book consists of an introductory Part I giving a general presentation of the specific kind of food ethics that is developed in this book. Part II consists of two chapters, the first being a critical analysis of the dominant scientific food regime of the industrialized countries. The criticism is directed towards the narrow and therefore also limited understanding of food and the consequences of the prevailing food production practices in most industrialized countries. Chapter Two shows how the development of modern biology as a natural science has turned into an intellectualization of the living and hence also of food. Chapter Three deals with the more practical implications of the intellectualization of food as they appear in everyday life. In Part III, I make a more constructive and philosophical response to the criticism. Chapter Four explores from a philosophical point of view the possibilities for consumers to relate ethically to food. This is followed by Chapter Five, which goes into more detail with regard to how food can be used as a way of relating to the environment, to others and to society. Chapter Six, the last one, presents the results from an empirical study on ethical traceability and discusses the issue of recognition in relation to ethical consumption. As mentioned, the literature on food ethics is still sparse and only a few philosophers have written on the subject of food. Apart from the "empirical" inspiration from consumer-supported agriculture and the empirical survey referred to in Chapter Six, my method has to a large extent consisted of exploring philosophical and ethical writings, which I believed would be fruitful from a food-ethical perspective – even though most of these philosophical writings are not about food at all. The idea is not to provide a critique of these philosophers' shortcomings but to use their work in a pragmatic way in order to elicit a new form of food ethics.

LIST OF TABLES AND FIGURES

ABOUT THE AUTHOR

Christian Coff (b. 1968) is Doctor of Philosophy and Research Director at the Centre for Ethics and Law, Copenhagen. He studied agricultural science at the Royal Veterinary and Agricultural University of Copenhagen (Denmark), Ecole Nationale Supérieure Agronomique Montpellier (France), Bundesanstalt für Züchtungsforschung and Technische Universität München (Germany). He is the founder of an organization for consumer-supported agriculture, called "Landbrugslauget", close to Copenhagen. He is currently leading several research projects on food, ethics and consumption. His academic interests include philosophical and ethical issues relating to food, consumption and environment.

PART I

FOOD AND ETHICS

CHAPTER ONE

EATING, SOCIETY AND ETHICS

Are the moral implications of food known? Is there a philosophy of nourishment?

Friedrich Nietzsche in *The Gay Science*

Cultures can be described by their values and morals, and the German philosopher Friedrich Nietzsche (1844–1900) surmised that there was a cultural link between food and ethics. He never analyses this link, however, and he leaves us with no more than a vague feeling that the intake of food must have some influence on human character and virtue – just as with the dietetics of the Greeks, as we shall see later on.

The question I pose is the opposite of Nietzsche's: not how does food influence our morals and values, but how do our morals and values influence our food consumption? The concept of "the political consumer" or "the ethical consumer" signifies that consumers are becoming increasingly committed to selecting goods from a political point of view. Well-known examples of this are the boycott of Shell as a response to its plans to sink an oil platform in the North Sea as a method of disposal, and the boycott of French wine as a retort to atomic bomb testing by the French military. This kind of political action effectuated by consumers clearly has an ethical foundation; in these cases, the protection of the environment and mankind.

The two cases received considerable attention in the media and were headline news items over a long stretch of time. Because they were dramatic, they were good media stories, but in most cases stories of production practices fail to gain any significant media exposure and thus remain hidden from consumers. This questions the actual effect of the politically conscious consumer.

For example, in 2001 the Nordic Council of Ministers published a report called *Food Labelling: Nordic Consumers' Proposals for Improvements*. This was the result of an attitudinal survey conducted among 1,300 Nordic consumers as store exit-interviews (usually as consumers were leaving supermarkets) and follow-up telephone interviews of the same people including some of the consumers' attitudes. A number of the results are very striking (see Table 1).

Table 1. Consumer attitudes. Selected results from the Nordic Council of Ministers (2001)

The Nordic Council of Ministers: Food Labelling: Nordic Consumers' Proposals for Improvements (TemaNord 2001: 573, 38)	
Question	*Result*
Do you think it is important to demonstrate your attitude as consumers, e.g. by boycotting products?	Close to 70% of Nordic consumers *think* it is important to demonstrate their attitudes as consumers, e.g. by boycotting products.
Are you ready to pay more for foods produced with respect for animal welfare and the environment?	Nearly three-quarters of the consumers *claim* that they are ready to pay more for foods produced with respect for animal welfare and the environment.
Do you prefer to buy organic commodities, even if they might often be a little more expensive?	Close to half of the consumers (48%) *claim* to prefer organic commodities, even if they are often a little more expensive.

Table 1 shows *the attitudes* of the consumers *when questioned about these issues*. This is what the consumers *think* is right *when they are asked* as part of a survey. But attitudes are one thing, reality is another. Attitudes in this area are more or less free, while real actions have their price: 48% claim to prefer organic food, but at present only 5–6% of the total food consumption in Denmark is organic. The consumption of coffee with "ethical labelling" subsumes an even lower percentage of the market.

The gap between what the surveyed consumers *think* and *do* is too vast to be an accident. The report from the survey makes it clear that "'political consumption' is far more common in attitudinal investigations than is reflected in real consumer behaviour".[2] Sociologists and anthropologists have known for more than three decades that people are *not* telling the truth about their consumption habits when asked about this in surveys.[3] This lack of coherence between attitude and action or behaviour arouses suspicion. Are the consumers *really* telling the truth about their attitudes? Or are they pretending to be more politically correct than they actually are?

It strikes me as highly improbable that the gap between attitudes and actual action and behaviour can be explained away as merely a matter of consumers bending the truth in order to *pretend* to be "politically correct" in

[2] Nordic Council of Ministers: *Food Labelling: Nordic Consumers' Proposals for Improvements*, p. 38.

[3] For this reason, some sociologists have developed a specific method, called "garbology" – the study of people's garbage, the assumption being that garbage does not lie. Garbage is solid empirical material.

the interviews. Attitudinal inconsistency with behaviour is not necessarily synonymous with exaggeration or pretentiousness. I believe that in order to *understand* the gap between attitudes and action or behaviour[4] a different approach must be taken.

Questions about consumption relate to consumers' self-awareness as well as to their social identity. For the most part, people *make up* pictures of themselves; they create their own identity by selecting elements from their own life story, often referred to as the "narrative identity", and from other sources, including ideas of how they *would like* to be. Thus, conceptions of ourselves are not necessarily realistic. They tend to be self-constructed and are partly illusory, yet they are important in social contexts and for identity. I suggest that this "constructed identity" should be interpreted ethically and that the gap between what consumers *claim* to do and what they *actually* do represents the gap between their vision of the good life and their real life.[5]

In this case, however, our object of concern is not merely "the good life" in general but more specifically "the good life" in relation to food production practices. We will be taking a close look at the consumers' vision of the good life in relation to food production practices. The survey suggests that it would not be too much to say that consumers, at least in the Nordic countries, have acquired a taste for ethics. One might safely assume that this is also the case in many other European countries.

The ethical interpretation of the gap between attitude and behaviour in attitudinal surveys of consumers suggests two things: first, that the ethical engagement of the consumers is high (and often underestimated, or rejected as a moral double standard due to the gap between attitude and behaviour); and second, that there must be some barriers that actuate this gap between attitude and behaviour and accordingly prevent consumers from putting their attitudes into practice. We are reminded of St Paul's famous words: "I do not understand what I do. For what I want to do I do not do, but what I hate I do . . . . For what I do is not the good I want to do; no, the evil I do not want to do – this I keep on doing".[6] With the reservation that eating is no evil, this seems to describe the situation of many consumers with a taste for ethics. There is the difference, however, that the consequences of the consumers' shopping is not evident, as the production history remains hidden from them. They may have an idea of what comprises "the good life" but not of the right action that leads to this.

[4] I do not distinguish here between action and behaviour, as is especially done in sociology. See, e.g. Kaj Ihmonen: "Sociology, Consumption and Routine". In Gronow, Jukka and Alan Warde (eds.): *Ordinary Consumption*, p. 10.

[5] Sometimes *attitude* is interpreted as referring to the identity as citizens, and *behaviour* as referring to the identity as consumers.

[6] Romans 7:15–19 (The Bible, New International Version).

1. THE INTIMACY OF EATING AND DIGESTION

Most people have an intimate relation to food and eating. We are more discriminating about food than about most other things. But why is this so?

The process of eating can be separated into several steps. Setting aside for the moment cultural and religious bias as well as all preconceptions of what is or is not edible, eating is often preceded by the view and the smell of the food. Our approach is often to use the visual aspect of the food to classify it as edible or inedible. This classification takes place at a relative distance. We then estimate the food by its smell. The sense of smell is of course far more intimate than that of sight, though not as intimate as that of taste. There is still some distance involved in smelling, whereas in tasting we are in close physical contact with the food in question. If we are not averse to either the sight or smell, we may then venture to go ahead and taste the food. Tasting is the final judgement on edible or inedible, the last gate the food has to pass where it still can be accepted or rejected. Swallowing is usually the point of no return.

Since the time of Aristotle the physiological sense of taste has generally been divided into four: salty, sour, bitter and sweet. From these basic elements all other tastes can be derived. This popular theory is very much consistent with other ancient Greek classification systems such as the four elements: earth, water, fire and air. But it tends towards simplification; many food researchers nowadays are convinced that a larger spectrum of tastes exists.

The gustatory sense is not isolated, but is related to the other senses. Most people who have caught a cold experience not only a temporary loss of the olfactory sense but also a reduction in the gustatory sense at the same time. It is also thanks to the olfactory sense that the first bite of, for instance, an apple tastes different from the next ones. Just as the olfactory sense quickly accustoms itself to new smells and becomes much less sensitive to them, so does the gustatory sense undergo a change based on this experience.

The gustatory sense is linked not only to the other senses but also to the mind, and, as mentioned, to cultural values. The eyes of newly caught seals are considered a delicacy by the Inuit. The Danish writer Jørgen Roos recalls in his *Recollections*:

> I was offered a special titbit, a freshly shot seal's eye. It is eaten by making a cut with a sharp knife and after that the contents are squeezed out in the mouth as if it were a large grape. One should remember to spit out the stone; it is the lens in the seal's eye.[7]

No doubt in most Western cultures the appealing eyes of the seal would be considered repulsive for eating purposes. Taste is further dependent on context, and not only the cultural context but also the state of mind and the actual atmos-

[7] Roos, Jørgen: *Recollections*, p. 5.

phere and situation in which the eating takes place. Moreover, there is clearly a subjective side to physiological taste. Neurobiology teaches that taste is memorized by the brain in the form of chemical signals, as specific markers or cellular and neural patterns. These memory patterns are altered by a series of parameters, some of which are social, emotional, biological, symbolic and so on. But this is not news in the world of gastronomy. The French hedonist and philosopher, Michel Onfray (b. 1959), writes:

> All gourmets know that part of the mysticism of the menu is due to the poetry of the words and that the title of the menu hides, disguises, covers up, reveals or lets us presume acts and artifices which have made possible the transformation of the natural product into its cultivated appearance.[8]

Digestion follows the tasting. The digestive tract is very versatile (more so than taste, I might add) and the human being can live off more kinds of food than any other creature. But we only eat a very small part of what is actually edible. Periods of hunger as in wars are well known for extending the perception of what is edible under duress.

The human digestion system starts already in the mouth. Bacteria and enzymes in the saliva initiate the decomposition of the food. Indeed, the actual cooking of food may be considered an extension of its digestion in that during the preparation the food in many cases is made more easily digestible. For instance, bread is made of cereals and like all other grasses these are not suited to human digestion. Cows can digest grass but human beings cannot. However, the fermentation of the dough, especially the lengthy fermentation, changes the structures of the nutrients in the bread and thus renders it more digestible to human beings. Saliva can also be used outside the body to prepare and conserve some kinds of food, as is the case when lightly chewed cabbage is stored in glasses, as is the case in some cultures. In the stomach the food is broken down with the aid of hydrochloric acid, bacteria and other microorganisms.

In the process of eating, elements of our physical surroundings are taken up and absorbed by the body. In the act of eating, the outer world becomes a part of the person who is eating, and in this sense, eating makes the outer involved with the inner. When food is swallowed, it does not immediately become a part of the body, but is rather placed in a specific cavity of the body, the digestive tract. This system, which in adults is about 12 m long and has a surface of about 400 m^2 (about the size of a tennis court), can be considered an *inner outside* – a channel or pipe through the human body. However, what is found in the digestive tract is not the human body itself, even though an element of it will later become part of the body. The digestive tract is full of various microorganisms; for instance, there are approximately 100,000 billion bacteria in the human body (compared with a mere 50,000 billion cells) dispersed over about 400

[8] Onfray, Michel: *La raison Gourmande. Philosophie du goût*, p. 54.

different human species. Many of the bacteria adhere to the cells in the gut and are indispensable to its function. The composition of microorganisms in the digestive tract is not stable but varies over time. The digestive tract of the unborn child is sterile, but during birth the child is inoculated with the microorganisms from the intestine and the milk of the mother. The composition of microorganisms in the digestive tract does of course have some general characteristics, but on the other hand it is just as individual as a fingerprint.

Present in the digestive tract, the content of which is not yet part of the human body, is another kind of life consisting of bacteria, fungi and other microorganisms that are essential for digestion. However, the digestion chambers are clearly not an "outer world" as we understand the concepts of surroundings or environment. They are a confined "outer world". The content of the digestive tract is in a borderland, and that makes it difficult to define: it is the bodily interior without being body and it is exterior without being surroundings. Eating is a transformation where what is "other" is transformed to the self. Eating confronts us with one of the most fundamental kinds of transformation, namely, encountering the *otherness* of our surroundings, which, through this activity of "internalizing", are transformed into oneself, into one's own body.

Our surroundings, which is nature, are *in-corporated* in order to be transformed through digestion (Greek: *pepsis*), and in the end to be incarnated (Latin: *carne*; meat), i.e. to become flesh – or perhaps some would say "reincarnated" (reborn), but now as a human being. During this in-corporation our surroundings are outer inwardness;[9] in the course of the digestion the outer world is transformed, and during this process it is therefore neither outer world nor body; finally, in the incarnation, our surroundings are embodied as a part of us.

In all living creatures openings to the outer world that allow it to enter the body are a necessity, but at the same time this renders the organisms vulnerable to hostile intruders like pathogenic bacteria. Many religious rites take into account that eating might expose human beings to danger. For instance, nuns in Greek Orthodox convents routinely bless themselves before and after each meal to protect themselves from dangers associated with the opening and closing of the body during the meal.[10] The outer world is far from being harmless. Letting food into the body incautiously is too risky, and precautionary measures must be taken to protect oneself. One must be both selective and particular about one's food.

The passage from one condition to another – as eating, which involves turning something external to the body into part of the body – is considered dangerous, because transitional states are neither one state nor the next. According to the English anthropologist Mary Douglas, they are in between and cannot be

[9] Schmidt, Lars-Henrik: *Smagens analytik*, (*Analytics of Taste*), p. 83.
[10] Lupton, Deborah: *Food, the Body and the Self*, p. 16.

defined; they are neither what they have been nor what they will become.[11] Digestion is one such transitional state or marginal period. Some kind of order is apparent at the beginning, because what we eat is classified as food. However, shortly after in the digestive tract the food enters the marginal period and is transformed into some indefinable substance. It cannot be described by some clear concept that would make us understand what it is. It seems to be complete disorder. Disorder is substance without borderlines and pattern but with the potential to turn into new and unforeseen states and therefore cause unpredictable changes in the future. Because of the dangers of the transitional states, rituals often surround transition, as the above example from the Greek Orthodox convent shows. Transitional rites are often symbolized as death and birth, and this symbolic or metaphoric language also covers eating. Food that is eaten was once living, but when it is eaten, it dies. However, we could also say that at the actual time of eating the food is usually already dead, and when we eat, it becomes alive again – but now as our bodies. The danger is of course that when we eat we are running the risk of being eaten instead, that the food for some reason is poisonous and in the worst case kills (or eats) us. The outer world would then not be reincarnated as our bodies but as itself, at the expense of our bodies.

The incarnation that follows the eating process led the German philosopher Ludwig Feuerbach (1804–1872) to put it this way: "Der Mensch ist was er ißt" (*Man is what he eats*).[12] Feuerbach wished to give a materialistic (and provocative!) answer to the anthropological question raised by Immanuel Kant approximately 50 years earlier: "Was ist der Mensch?" (*What is Man?*). Feuerbach had no intention of hiding his own very materialistic point of view; instead he made a virtue of it. The point of departure of his extremely materialistic philosophy was the burgeoning food science, a relatively new scientific area at that time. Comparisons of the chemical compositions of the food with the content of the human body showed many points of similarity, and this gave Feuerbach the inspiration for his radical conclusion. Positive science had answered Kant's anthropological question. And the resemblances are striking: too little iron intake can be measured in the blood as lack of iron; one becomes fat from eating fat foodstuffs, one becomes thin from eating lean food; one becomes healthy from healthy food and happy from funtex; and if one eats carrots . . .! No, this is quite absurd – we all know that we do not become what we eat. The similarities can only be found at a molecular level and even there only in very restricted areas. Human beings are substantially different from what they eat. The digestion and decomposition of food and the synthesis of new molecules in the human body make it reasonable to talk about molecular "transubstantiation", i.e. transformed substance.

[11] Douglas, Mary: *Purity and Danger: An Analysis of Concepts of Pollution and Taboo*, pp. 95–98.
[12] Feuerbach, Ludwig: *Die Naturwissenschaft und die Revolution*, p. 367.

This view is also taken by the German-American philosopher Hans Jonas (1902–1993). In his book *Das Prinzip Leben* Jonas presents, with his own terminology, a biological philosophy or philosophy of organism.[13] Central to this philosophy is the ontological status of the living: what is "being"? Traditionally, we think of the substantial and material composition of the living as essential qualities. But Jonas takes the opposite position:

> Organisms are never identical from a substantial point of view, but nevertheless the self of the organisms remain a "self", precisely because they are not made of the same matter.[14]

If an organism turns out to be substantially identical at two different times, there can only be one reason: it is no longer alive. Metabolism, the flow of substance "through" the organism, is essential for survival and continued identity. In that sense, living organisms differ from their substance and therefore also from what they eat. Describing the substantial aspect of organisms as accidental, Jonas believes that *form* is a much more characteristic quality of organisms. In biological theory the concept of form is closely related to the concept of *organization*, because it is the organization of the matter that gives the substance its form. Form or organization is transcendent, as it remains more or less the same and keeps its identity in spite of the metabolic replacement of substance in the organism. Jonas goes even further when he writes:

> [T]he form is not the result of material composition but the cause of it . . . . 'Selfness' (German: *Selbstheit*), as long as it lasts, is the continuous renewal carried by a flow of that which always wants to be the other. It is only with the active 'self-integration' of life that the concept of 'individuality' acquires its ontological status, in contrast to the phenomenological concept of individuality. This ontological individual, with its existence through every moment, its duration and 'selfness' in the duration, is first and foremost the function of the individual itself; it is its own interest and its own continued performance. In this self-preservation the organism relates in two ways to its own substance: matter is substantial as matter and accidental to the form. The organism merges with the material composition of the moment but is not tied to the material composition at any specific time, only to the form that it constitutes in itself, dependent on the matter as substance but independent of the 'selfness' of the matter; the functional identity of the organism does not merge with the substantial identity. In short: the relationship of the organic form to the matter can be described as an urge to freedom (*bedürftiger Freiheit*).[15]

The form remains, whereas the substance is replaced. Jonas makes of form the essence of life and of substance, accidence. Form is described as a succession of moments marked by differences in substance. *Temporality* becomes an important

[13] The book *Das Prinzip Leben* was originally published in 1973 as *Organismus und Freiheit*, which was a rewritten and expanded version of *The Phenomenon of Life*, 1963, New York. Quotations are here translated from the German book *Das Prinzip Leben*.

[14] Jonas, Hans: *Das Prinzip Leben*, p. 145.

[15] Ibid. pp. 149–150.

characteristic of the living in the biological philosophy of Jonas. The relationship of form to matter is not passive; rather the form is actively shaping the matter. The form is independent of the actual matter, though of course dependent on there being matter at all. Jonas describes this as "dependent independence". This dialectical freedom of the form is not absolute but limited by the actual need of matter. It is also a "forced freedom", as the form must necessarily replace the matter through the metabolism. Dialectical freedom is a fundamental quality of life.

Living beings, including the human being, must eat to stay alive: what is eaten is the world. The need for nourishment forces organisms to open up to the outside world and to develop senses orientated towards the outer world. The openness in the meeting with the world makes the experience of the world possible. Jonas describes the openness as the transcendence of hunger. It is the lack of food that makes organisms reach for the world, to grasp and sense the world: to have a world is the transcendence of life. The unfamiliar and alien surroundings could potentially become a part of oneself and at the same time one could become a part of the surroundings. The reaching out for the world is the basis of *intentionality* as the fundamental quality of life.

Today, if "Der Mensch ist was er ißt" (*Man is what he eats*) is to have any meaning, it is not from a purely materialistic point of view. Matter is accidental, whereas the form or organization regulates the metabolism and confers individuality on the organism. Eating a carrot does not turn us into a carrot; the substance of the carrot is organized into another form. It is incorporated into the human body and thus given the form of a human being. However, Feuerbach's aphorism continues to carry a meaning and to make sense to many people. Otherwise it would not have been quoted as often as is still the case. Eating is discriminatory; when we eat, we are generally particular about our food. However, we choose food not only from a materialistic point of view, as is the case when considering the healthiness of the food for the body. Just as important for the discriminating regimes are cultural values and the fact that food is related to identity and self-understanding. Man is what he eats, not only in a substantial sense but also in a cultural sense.[16]

2. EATING IN BETWEEN LIFE AND DEATH

Eating is a continued holocaust (from Greek: *burnt offering*), as Michel Onfray remarks: to exist the body needs to take in corpses, things torn up by the roots, ripped out of their natural environment, food that is fermented and putrefied and hanged meat.[17] It does not sound very delicate but these are nonetheless some of

[16] See, e.g. Baudrillard, Jean: *La société de consommation* and Bourdieu, Pierre: *Distinction. A Social Critique of the Judgement of Taste.*

[17] Onfray, Michel: *La raison Gourmande. Philosophie du goût*, p. 111.

the ways that we treat our food before eating it – a smoked ham, for instance. However, since the Second World War the concept of holocaust has acquired another meaning, with the concentration camps reminding us of modern, industrialized meat production, not least in the frightening rationality that calculates the most effective mass slaughter that leaves no trace. And, as with the holocaust, the public in general do not wish to know about the work of the slaughterhouse, even though it puts food on their table. Mass slaughter of animals exceeds what most people can bear – but of course that does not make it disappear.

This is the tragedy of eating. Eating always implies sacrificing something, eating must always have a victim, and there is always something or someone who has to die when others eat. Hunger leads to the "dilemma of eating": when we eat we kill – but if we do not eat, we will die. Whether we eat or not, it is going to be the death of some. Hunger leads to death either for us or for others. Not to eat is to let others eat oneself; it is to give life with one's own life in the grave. The aesthetics of gastronomy serves to hide the banal necessity and the barbaric compulsion of eating. Aesthetics turns the necessity of eating into a pleasure unsubordinated to anything else, i.e. pleasure without purpose, as the purpose is not satiation but pleasure itself, which in this case is the aesthetic satisfaction gained from the meal.

Food is organic; it comes from nature. Only a small part is inorganic, like salt. Organic material is living – or at least it has been. The food we eat has had a life and is now dead or is going to die. Some food is eaten alive, some eaten dead. Animals are usually eaten dead and also bought dead and butchered. Plants are eaten both dead and alive (e.g. vegetables and fruits continue their metabolism after having been separated from the soil, from the rest of the plant, etc.). We consider it a horrible violation of the animal to eat it alive, as it is able to feel pain. It is humane to kill the animal before eating it.

Some, though, find it offensive to eat meat at all. For them eating meat is equated with murder, as formulated for instance already in 1791 by the vegetarian John Oswald in his book *The Cry of Nature; Or, an Appeal to Mercy and to Justice, on Behalf of the Persecuted Animals.*[18] Eating habits often reveal different ways of relating to, and thinking about, nature. The debate is not new. One of its oldest sources goes back to a dispute on vegetarianism in 450 BCE, when Ovid (43 BCE–17 CE) took issue with the earlier views of Empedocles (*fl.* 450 BCE). The consumption of meat was even seen as a killing of relatives; since everything comes from the earth and returns to earth again, we will inevitably eat one other. It is unclear, however, why this reasoning only applies to meat; it seems that the same argument serves just as well for plants.

[18] Oswald, John: *Radical Food. The Culture and Politics of Eating and Drinking 1790–1820*, pp. 143–170.

Is the consumption of meat, i.e. the incorporation of the animal into oneself, a violation of that animal? Some think not. They regard eating as a negation of death, and eating even becomes an appraisal of what is eaten. When the French philosopher Jacques Derrida (1930–2004) was asked about the moral implications of the meal, he simply replied that the moral question has never been whether one should eat or not, dead or alive, human or animal; the essential premise is that we *have* to eat and eating is thus in itself good and just. For Derrida, the killing of the animal is of no concern – since we have to eat to stay alive. Eating is thus a rejection of death, and the moral question of eating is concerned with the "good" meal. Doubtless Derrida delights in the famous French hedonistic tradition of the culinary arts, which is centred on the pleasures of the table; talking about the origin of the food and the animals eaten is not only impolite and indelicate but for the true hedonist it is also unthinkable. It is separated off and carefully eliminated; there are no words for it in the hedonistic food regime. Rather than consider the former well-being of the animals eaten, the hedonistic tradition concentrates on the present and the future. Eating means the death of some; that is how nature works and we can do nothing about it. But more importantly, it is the joyful continuation of life – not of course for those eaten but for those who eat. The hedonistic understanding of the good meal is more about aesthetics than ethics. This is how aesthetics usurps ethics,[19] and this is particularly evident within gastronomy.

The German performance artist Daniel Spoerri, who often incorporated food in his art and happenings, thought of eating and being eaten as an expression of an almost metaphysic metabolism: the continuous reappearance of the same. Eating is the continuation of life in another form.

These are the poles between which eating takes place and which also cannot be separated. Death and life always go together in eating. Eating meat can be seen as a killing, or it can be considered a necessary but beautiful metabolism where life gives life.

3. THE SOCIAL MEANING OF THE MEAL

In his short essay *Sociology of the Meal* (1910) the German sociologist Georg Simmel (1858–1918) reflects on the meal and how it is linked to the tension between individuality and community. Simmel starts by noting that first of all eating is both common and extremely individual:

> Hence, of all the things that people have in common, the most common is that they must eat and drink. And precisely this, in a remarkable way, is the most

[19] See Korthals, Michiel: "Taking Consumers Seriously: Two Concepts of Consumer Sovereignty", p. 208.

> egoistical thing, indeed the one most absolutely and immediately confined to the individual. What I think, I can communicate to others; what I see, I can let them see; what I say can be heard by hundreds of others – but what a single individual eats can under no circumstances be eaten by another. In none of the higher spheres is it the case that others have to forego absolutely that which one person should have.[20]

Eating is individualistic, as the same food cannot be eaten twice; food is ultimately used and transformed during eating and digestion and cannot also be eaten by another person (unless we eat each other!).

However, during the shared meal individuality is transgressed and elevated into community and social interaction. Simmel writes:

> Communal eating and drinking, which can even transform a mortal enemy into a friend for the Arab, unleashes an immense socializing power that allows us to overlook that one is not eating and drinking "the same thing" at all, but rather totally exclusive portions, and gives rise to the primitive notion that one is thereby creating common flesh and blood.[21]

Holy Communion, Simmel emphasizes, is the ritual in which the suspension of individuality finds its most sublime form and symbolization. During Holy Communion, where the bread and wine symbolize the life of Jesus, it is not the case that one cannot eat what has just been eaten by another; everybody eats the same; everybody receives a part of the unbroken whole.

In its most simplified conception, the social meal also creates the same flesh and blood, as mentioned above. Ludwig Feuerbach would have said that if two persons eat the same they also become the same, but in a less materialistic interpretation "becoming the same flesh and blood" is a metaphor from genealogy. The ties of blood refer in this case to the ties among the individuals in a community. The social meal rejects egoism and binds individuals together in a community and a common identity. The egoistic satisfaction of the fundamental need for food is ignored and instead the food is shared; one does not eat what the others are eating (or could be eating) and one certainly does not eat the others!

Thus, even though eating is basically egoistic, the common meal has the power to override this egoism, and more than anything else it possesses the ability to institute solidarity and community feeling. Of all the things that human beings have in common, eating is among the most fundamental. Eating is the lowest and most egoistic common denominator, and this is why we find in the meal the best-suited instrument of instituting a community; the meal does not demand specific qualifications, everybody can participate without any other conditions or competences than being hungry. The meal appeals to everybody and this renders it sublime in overruling egoism among all individuals. Simmel notes that if eating had not been low enough to include everybody, the rise to

[20] Simmel, Georg: "Sociology of the Meal", p. 130.
[21] Ibid. p. 131.

sophisticated social conventions – as in the aesthetics of the culinary art – would not have been possible. The origins of the social meal and the solidarity it gives rise to are thus not to be found in a common taste. This arises from the widespread and common hunger, not an egoistic hunger but a socialized and disciplined hunger, hunger transformed from a natural and bestial instinct into civilized and cultivated manners. Community is founded in the taming of hunger, which is the civilizing process.

Individuality and sociality are reconciled in the meal; people eating together enter into a contract or a pact. The change in the aesthetic conventions of the dining table at the beginning of the 16th century was also the start of a reorientation of table manners. Simmel points to the fact that as the meal becomes a social matter, it is organized into a more aesthetic and stylized form. A conclusion he draws from his observations of lower social classes, where the material sense of the food dominates in contrast to the higher social groups where the meal is dominated by codes of rules and behaviour regulations. The plate and the glass symbolize the order balancing the relationship between egoistic individuality and solidarity; the plate and the glass ensure that everybody gets their share of the whole and at the same time they set limits for how much can be taken (as one cannot take more than there is room for on the plate or in the glass). It is a socialized distancing to the animal in the human being and the lowest common denominator. According to Simmel, the aesthetics of the meal serves to suspend the individual desire for satisfaction, and accordingly it is important to moderate or temper the individual character of the dining table. In his view, accentuating the individual character of a course or the aesthetic arrangement of a table is indecent and bestial. Simmel dissociates himself from the individualization of the course as he considers it akin to cannibalism.[22]

Food and the meal are for these reasons especially suitable as symbols of solidarity and community feeling. Simmel praises the common, civilized meal for its ability to elevate the physiological primitiveness of eating to higher social forms and social interaction. Leslie Goften also touches upon this when he writes:

> [F]ood not only symbolises cultural values, but also forms a medium through which social relationships are expressed, from the intimate, face-to-face relations within the family, to the relations between regions and nations.[23]

Food is used to express social relations in various ways. In Arctic Inuit societies it is considered impolite to offer thanks for a meal. As a natural custom they always divide the food that the hunters bring home between themselves. As the Arctic explorer Peter Freuchen was thanking the Inuit for meat given to him, an old man promptly corrected him:

[22] Simmel is not very clear about his dissociation from anthropophagy, which seems only to be used to make a parallelism. In spite of the fact that Simmel views the meal as a basic means of institutionalizing community, he does not see it as a social pact of the meal not to eat each other.

[23] Goften, Leslie: "Bread to Biotechnology; Cultural Aspects of Food Ethics", pp. 121–122.

> You must not thank anyone for your meat; it is your right to receive parts of it. In this country, nobody wishes to be dependent on others. Therefore, there is nobody who gives or gets gifts, for you thereby become dependent. With gifts you make slaves, just as with whips you make dogs.[24]

One does not need to thank anyone for the meal, as everybody has a right to a share in it, independent of who provided it. Thanking anyone for the meat is a sign of a lack of solidarity: the Inuit do not thank the hunter, instead they celebrate his skills.

In some languages we find a double meaning of the word "course" – as in Danish *ret*, Swedish *rätt*, Norwegian *rett* and German *Gericht*. In these languages the same word has both a gastronomic sense, as in "course" or "dish", and a legal sense, as in "law" or "justice". The etymological root of these words is thought to be in the old German word *rextia*, which means to straighten or to make even. The development of this word into its gastronomic sense is believed to have been influenced by the Low German "richte", which has the same etymological roots.

This allows us to assume a connection between the juridical and gastronomic meaning of the word: the fundamental right in some cultures may have been the right to food. This is easily understandable, since the denial of food equalled exclusion or at least the inferiority of the individual. Distribution, especially the distribution of food, is a fundamental dimension of justice in any society. The right of all members of the community is to have their share of the available food, even if they have not earned it or deserved it. With Simmel in mind we can say that the right to food is the basic constitutive element of a community.

4. FOOD AND ETHICS IN HISTORY

There is a long tradition of relating food and eating to ethics and morals. In most cultures we find *food regimes*, i.e. foods that are surrounded by taboos, specific meanings and values, and also very precise descriptions of what to eat and how to eat. The following is a short history of the relationship between food and ethics.

In antiquity this relationship was expressed in dietetics (from Greek: *diaité*). The French philosopher Michel Foucault (1926–1984) in his book *The History of Sexuality* refers to Hippocrates, who described how in the beginning the food of human beings was like that of the animals and how human beings in the course of time distanced themselves from this natural way of eating things raw by preparing the food before eating it.[25] Thus, there is no doubt that Hippocrates understood food not only as natural but also indeed as *cultivated nature*. Dietetics related food to bodily health; the diet was the right food essential for

[24] Freuchen, Peter: *Book of the Eskimo*, p. 154.
[25] Foucault, Michel: *The History of Sexuality*. Vol. 2, p. 99.

the cure of the patient. However, the dietetics of the Greeks was not just used as medical treatment:

> '[D]iet' itself – regimen – was a fundamental category through which human behaviour could be conceptualized. It characterized the way in which one managed one's existence, and it enabled a set of rules to be affixed to conduct; it was a mode of problematization of behaviour that was indexed to a nature which had to be preserved and to which it was right to conform. Regimen was a whole art of living.[26]

Ancient Greek regimen was not only about the right nourishment of the body but was just as much a question about the conduct of life and human existence. For this reason dietetics was also a reflection on the good life. The right diet and the right conduct of life could not be separated. Foucault's analysis of the dietetics of especially Hippocrates and Plato points to the twofold meaning of this food regime: the healthy body and the healthy soul. Dietetics was the art of living. Dietetics was ethical when it showed a way of living the good life: avoid extremes and eat with moderation (temperance). Dietetics was a balancing between too much and too little; it was the care for one's own body and the self. Dietetics was a specific way of constituting oneself as a corporal and moral subject; it was a *technology of the self.*

It is almost an everyday experience that food has a physiological impact on the body and that some kinds of food can lead to tiredness, excitement, dizziness and many other states. Lack of specific nutrients, as in the case of malnutrition, also affects the body and the mental state and can lead to depression, anxiety, irritability, emotional lability, hysteria, etc.[27] Therefore, the practice of regimen is an art of living; it is a strategy involving the body and the elements proposing to equip the individual for a rational mode of behaviour.[28]

The books of Deuteronomy and Leviticus in the Old Testament contain extensive examples of religious dietary rules that prescribe what is allowed to be eaten and what is forbidden. In Deut. 14 we read: "Do not eat any detestable thing. These are the animals you may eat; the ox, the sheep . . . you may not eat the camel, the rabbit . . .". There are many interpretations of the meaning of these rules. Mary Douglas in her book *Purity and Danger* (Chapter Three: *The Abominations of Leviticus*) gives a brilliant review of the most common interpretations of the meaning of the dietary rules.

Some readers have taken a modern pragmatic view, considering them as an expression of food sanitation. The text of Deuteronomy should in this interpretation be comparable to modern food security or safety rules. This is unlikely to be the case, however, as there is no indication of this at all in the Bible.

[26] Ibid. p. 101.
[27] Benton, David: *Food for Thought*, p. 17.
[28] Foucault, Michel: *The History of Sexuality*. Vol. 2, p. 108.

In the second interpretation it is stated that the objectives of the dietary rules were self-control and prevention of thoughtless action and injustice. The dietary rules were a means of achieving the good life. This ethical interpretation has its origin in the periods of Alexander the Great and Hellenic culture.

A third interpretation is the view that the dietary rules should function as a protection from foreign influence. In her book, though, Douglas is rather swift in rejecting this as irrelevant and she does not see such protection as a mode of conserving cultural identity. Dietary rules prescribe specific eating habits and thus also a specific conduct of life that is important not only for cultural reasons but just as much for the understanding of the individual's identity. Douglas also proposes a fourth religious interpretation:

> [T]he dietary laws would have been like signs which at every turn inspired meditation on the oneness, purity and completeness of God. By rules of avoidance holiness was given a physical expression in every encounter with the animal kingdom and at every meal. Observance of the dietary rules would thus have been a meaningful part of the great liturgical act of recognition and worship which culminated in the sacrifice in the Temple.[29]

What Douglas is indicating here by referring to the dietary laws as *signs* that inspire meditation actually comes close to the kind of food ethics that will be developed in Part III of this book. However, I intend to follow a different path from Douglas. First, the context in this book is not religious, but ethical. Second, I will not speak of signs but of traces. At the end of this chapter and in Part III I will examine the philosophical consequences of regarding food as a trace of the production history. It is argued that the ethical dimension of food indeed involves looking at how the food is produced, i.e. the specific food production practices.

According to the Dutch philosopher Hub Zwart (b. 1960), the moral implications of the dietary rules are abandoned in the Gospels, i.e. in Christianity.[30] In the light of the interpretations presented by Douglas we can take the case even further and say that when dietary rules are abandoned in Christianity, this also implies that the other interpretations mentioned above would lose their meaning. Food would no longer be related to cultural identity. It is an important aspect of the Christian tradition that food loses its relation to self-understanding and can no longer be used as a sign that inspires meditation of the holy. The Gospel of Matthew can be seen as a direct refutation of the Old Testament, when Jesus says:

> What goes into a man's mouth does not make him 'unclean', but what comes out of his mouth, that is what makes him 'unclean'.[31]

[29] Douglas, Mary: *Purity and Danger*, p. 58.
[30] Zwart, Hub: "A Short History of Food Ethics", p. 118.
[31] Matthew 15: 10–12.

As the disciples do not understand, they ask him to explain. And he answers:

> Are you still so dull? . . . Don't you see that whatever enters the mouth goes into the stomach and then out of the body? But the things that come out of the mouth come from the heart, and these make a man 'unclean'. For out of the heart come evil thoughts, murder, adultery, sexual immorality, theft, false testimony, slander. These are what make a man 'unclean'; but eating with unwashed hands does not make him 'unclean'.[32]

If the dietary rules formerly had a hygienic purpose, this is invalidated by Jesus who says that unwashed hands do not defile the man, and that it is of no moral importance what one eats. The nourishment becomes insignificant; what really counts is the food for the mind: "Do not work for food that spoils, but for food that endures to eternal life, which the Son of Man will give you".[33] According to Zwart there is nothing moral in the relation to food in Christianity; on the contrary, Jesus encourages averting one's eyes from the earthly and material world.

I consider this interpretation too radical, however. Another interpretation would be that as food in the Old Testament is given a central position, Jesus in this context is settling up with the traditions of the Old Testament. Thus the message is: do not exaggerate the traditions of eating habits as they may turn the focus away from what really counts, which is faith.

Yet in spite of this interpretation, food in the New Testament is also used as a picture for what is important – as a sign that should remind us of God. The earthly food should lead the thoughts of the eating person towards holiness:

> And he took bread, gave thanks and broke it, and gave it to them, saying, "This is my body given for you; do this in remembrance of me". In the same way, after the supper he took the cup, saying, "This cup is the new covenant in my blood, which is poured out for you".[34]

This text from Luke constitutes the background for the Holy Communion. The theophagy of the Holy Communion, where the flesh of Christ is symbolized by a wafer and his blood by wine, is one of the biggest mysteries of Christianity: the presence of Christ as bread and wine. The scholastic term *transubstantiation* refers to the metamorphosis occurring in the bread and the wine during Holy Communion. It is substance that trans-substantiates, transforms into a new substance. This theophagy, which Christianity shares with many other religions and which also existed in antiquity, must be based on a killing.[35] The flesh of Christ is eaten in the form of bread and wine during the symbolic meal of Holy Communion; the corpus of Jesus Christ is incorporated into the body of the

[32] Ibid. 15: 17–20.
[33] John 6: 27.
[34] Luke 22: 19–20.
[35] Böhme, Hartmut: "Transsubstantiation und symbolisches Mahl. Die Mysterien des Essens und Naturphilosophie", pp. 139–140.

Christian. The external world is consumed and transformed to "inwardness" to become faith. This shows how strong a metaphor eating is for the appropriation of the world; and how nature is transformed into culture.

For the moment let us leave the Gospels and take a huge jump in time to the end of the 17th century (I will return to the intervening periods in Chapter Two). Food supply on a societal level becomes an important issue in the public debate during this period. Malthus (1766–1834) in his famous work *An Essay on the Principle of Population* (1798) describes how food production is unable to follow the growth of the population and that this inevitably leads to starvation. According to Malthus, nothing could be done to prevent it, as this rule was a natural law.

At present, lack of food and famine are increasingly being submitted to debate in a more or less scientific manner. However, some are beginning to consider famine not as a natural law with severely local consequences but as a problem that the state could and should take care of. Famine is transformed from a natural law into a cultural and moral problem. In *The Threepenny Opera* by Bertolt Brecht (1898–1956), Macheath (Mack the Knife) and Jenny from the criminal underworld sing the ballad *What Keeps Mankind Alive?*:

> You gentlemen who think you have a mission
> To purge us of the seven deadly sins
> Should first sort out the basic food position
> Then start your preaching, that's where it begins
> . . .
> So first make sure that those who now are starving
> Get proper helpings when we do the carving.[36]

The famous aphorism from the ballad states: "Food is the first thing, morals follow on".[37]

Starvation overrules the morality of the ravenous because it threatens death and destruction. To escape alive from starvation the ravenous will not hesitate to violate their own or society's morality. Starvation leads to amoral and antisocial acts. The medieval Swiss doctor, Paracelsus (1493–1541), put it thus: "All that live must eat and all that is eaten must be digested; from this digestion departs philosophy".

In western industrialized countries it is hard to imagine what starving is like; the surplus production of agriculture has reduced starvation to such an extent that the World Health Organization (WHO) has declared corpulence to be the bigger threat to mankind. More people die from being too fat than from not getting enough to eat. In the Western world we can try not to eat, as we do when we go on a fast, but this is incomparable to enforced hunger. We are therefore

[36] Brecht, Bertolt: *The Threepenny Opera*, Second Threepenny Finale, p. 55.
[37] "Erst kommt das Fressen, dann kommt die Moral", p. 55.

no longer familiar with starvation, and are unable to say whether "Food is the first thing, morals follow on". Of course, aphorisms push things to extremes. Mack the Knife and Jenny claim that to survive the poor must not merely compromise their own moral values but actually leave them entirely out of account. Only when man is not starving has he the *possibility* of acting morally; and even then satiety is no guarantee of moral and justifiable acts. What happens, though, is that satiety allows ethics to become vocal, it creates a space for a discourse on ethics, and it allows ethics to become a subject in the debate, whereas hunger excludes food ethics. And the first obligation of the satisfied is to assure that nobody is starving, to secure a reasonable and just distribution of food. Where starvation sets aside and neglects ethics, the distribution of food to all is the *course of justice*. This is to put ethics on the agenda.

So far, most ethical regulations pertaining to food production have been put on the agenda by authorities on a national or European level (often in collaboration with interest organizations). Food ethics is becoming part of the agenda, as can be seen in a survey of the many new councils dealing with this specific topic. In 1998 the *Food Ethics Council* was founded in the United Kingdom; in 1999 the *European Society for Agricultural and Food Ethics* was established; in 2000 the *Dutch Platform for Agricultural and Food Ethics* was set up, while the Food and Agriculture Organization (FAO) of the United Nations has created a *Panel of Eminent Experts on Ethics in Food and Agriculture*. The overall picture, drawn from these more or less official sources, is that food ethics deal with moral and ethical questions related to food. A common consensus also seems to exist that food ethics emanate from a general concern about the development in food production practices. This concern is directed towards the health of the consumers, food security, the autonomy of the consumers, consumer information, hunger and starvation in the third world, sustainable development in rural areas, equal participation in decision-making, the responsibility for future generations, animal welfare, protection of the natural resources and the biological diversity, ethical research and, finally, the development of an ethical codex or guidelines for food production practices. These are worthy ideals but difficult to live up to. It is not too much to say that food ethics bite off more than they can chew. The figure of 800 million starving people in the world reminds us of that.

5. FOOD ETHICS AND THE PRODUCTION HISTORY

Compounding food and ethics into the single concept of food ethics is a recent innovation. At first sight it seems to be an oxymoron, i.e. a concept constructed from two contradictory terms: ethics excludes food and food excludes ethics. Some, especially within the hedonist tradition, might even see food ethics as a result of bad taste, likely to spoil the appetite and the good meal. Food ethics are probably not going to be a suitable dinner-table subject in the near future. There is no desire to spoil the consumers' appetite, even though that is very likely to come about when we consider today's food production practices. Rather, the intention of food ethics is to discuss and formulate visions of the good life in food production practices and thereby increase the pleasure of the meal.

The emergence of a new concept is not accidental; it reflects new directions in the thinking and consciousness in the society. The concept of food ethics is used to designate a wide field of new concerns about food production practices. Food ethics, as they are understood to today, certainly differ from the understanding of the relationship between food and ethics in earlier times. Food ethics are not only about health and the maintenance of personal virtues, centred on the constitution of one's own human body and self. Food ethics are today also about the care for others. Food ethics as developed here include the ethics of the environment, the society and others in so far as these are related to food.

It is no exaggeration to claim that consumers have acquired a taste for ethics. So far I have mostly concentrated on food and its consumption. Before proceeding in Part III to a deeper philosophical analysis of the consumers' taste for food ethics, I also need to say something about ethics and their relation to food.

The aim of ethics is *praxis*. According to Aristotle (384–322 BCE), in his *Nicomachean Ethics*, ethics is "striving for or aiming at the good". What, though, is the good? And what is the good act? This is what ethics involves. Accordingly, within the realm of ethics, we must imagine the good in order to determine what is worth striving for. It is the vision of the good life that succeeds. However, this vision should not only be imaginable but also be convertible into *action*.

Actions always take place in specific situations. We cannot imagine an action without its setting. These situations are characterized by openness, inasmuch as different people are free to react very differently to the same situation. It is possible to take different kinds of action in a given situation: good, less good and bad. Some situations seem to be very similar, but they are never entirely identical. They always differ in some respects. Thus, the good action becomes dependent on the actual context. The point of reference for ethical reflection is twofold: it is the generalized or universal vision of the good life, on the one hand, and the employment of this vision, for instance, in the form of guidelines for real life, on the other. It is an intermediary between the generality of the vision and the specificity of the actual situation. The general ethical reflections must be adapted to the actual situation in order to be able to function as guidelines for acting in the specific situation.[38]

[38] This illuminates the difference between ethics and norms. Ethics are reasoned or substantiated, whereas norms are more or less non-reflected guidelines for acting, formulated as "you must . . ." or "you must not . . .", without any reference to a reason or the specific situation.

The second important lesson from understanding actions always taking place in the context of situations is that *one is not alone*. To be in a situation is precisely to be in a specific position in relation to something else, to other persons, living beings or things. The vision of the good life must necessarily include others, for life is never isolated from others but is always lived with them. Actions do not take place in isolation but always include others. Ethics comes into existence in the presence of the "other", or as Emmanuel Lévinas (1906–1995) put it: "We name this calling into question of my spontaneity by the presence of the Other ethics".[39]

The good life can not *only* be my own good life but it includes that of others, since my own good life is unthinkable without the other and his or her good life. My own good life includes the good life of the other. Thus, I must also be able to imagine or think of the good life of the others by putting myself in their position. The relatedness of ethics inspired the French philosopher Paul Ricœur (1937–2005) to state that ethics are the vision of the good life with and for others.[40]

However, Ricœur does not leave it at that. For we not only live in the context of face-to-face relations but are also part of a larger society, where justice is supposed to provide equality and fairness. In society, justice is institutionalized. Thus, Ricœur describes ethics as the vision of the good life with and for others in just institutions. Food production is one such institution in society and we may therefore conceive of food ethics as *the vision of the good life with and for others in fair food production practices*.

We must now ask who, within the sphere of food ethics, are those *others?* Let us imagine the consumers in the store, confronted with the presence of the food. They may with good reason ask what this has to do with ethics. For those consumers who know nothing about the practices of food production, there can be no food ethics. Food is ethically irrelevant to them. In the shops food is presented as "dead life", for which no moral claims can be made with respect to the life it once was. For this food we cannot harbour any ethics. We are likely to have a taste and an aesthetic opinion, but no ethical concern can be expressed. This is also evident when we consider that we cannot violate food in itself. Living beings can be violated, but food cannot.

Why is it then considered unethical to sell food with pathogenic microbes? It cannot be because it is a pity for the food, since the food is "dead", anyway. However, it is a pity for, and maybe even a violation of, the consumers who are going to eat the food and consequently get sick. Why do some consumers consider it unethical to eat eggs from battery hens? It is not because it is a violation of

[39] Lévinas, Emmanuel: *Totality and Infinity*, p. 43.

[40] Ricœur, Paul: *Oneself as Another*, chs 7, 8 and 9. Instead of "the vision of" Ricœur uses the words "aiming at", which are also used by Aristotle in his definition of ethics (aiming at, or striving for, the good life) in his *The Nicomachean Ethics*.

the eggs to eat them; nor is it because they were laid in a battery. Rather, they consider it unethical because it is a violation of the battery hens – and perhaps the quality of the eggs is deteriorated in comparison with more "natural" eggs, making them less attractive to the consumers. Why does the tasteless mixture of different high-quality food ingredients seem to be a pity and a waste to the eyes of a cook? It is not out of pity for the ingredients but because it is disdainful of the work put into the production of the food ingredients and an insult to the people who could have enjoyed a good meal.

In this light, therefore, food ethics do *not* concern the end product, the ready-to-eat food.[41] Instead, food ethics concern the production practices: *we know that food has its origin in the living world and that there must necessarily be a production history*. Even if we know nothing about the production practices when we buy the food, we still know that there must be a history and that this history can be cruel and ugly, as in a brutal slaughtering not respecting animal welfare, or it can be good and beautiful, as in the cases where the animals are treated properly and with respect for their natural behaviour. Knowledge of the production history of the food constitutes the basis for the ethical standpoint of the consumer. *The production history forms the basis of consumers' food ethics*. Without knowledge, consumers are excluded from acting (or shopping) ethically. They cannot ask the question: could the production history be different?

The case of genetically modified food illustrates this very well. Many Europeans oppose genetically modified food and wish to avoid it. But this is not always a simple task, as genetically modified plants can easily spread beyond control, e.g. as pollen is spread from one field to another or as the result of mixing genetically modified with non-modified plants during the processing (usually there is no visible distinction between genetically modified plants and conventional plants). The European authorities in this field, the European Union (EU) Commission, have proposed a labelling procedure stating that only if the content of genetically modified organisms (GMOs) exceeds a specific limit does it have to be labelled. This makes it completely impossible for the consumers to avoid genetically modified food. Being "forced" to eat, or being unable to avoid, genetically modified food, opponents can soon be expected to retire from the fray. This makes it impossible for the consumers to express their ethical engagement in food.

Food ethics stretch back towards the past, towards the origin of the food: towards the living nature that the food comes from and towards the production practices under which the food has been produced and distributed to the consumers. Food ethics is about the ethics of food production practices, all the way from breeding and farming to food processing and distribution.

[41] *Fair* distribution of food, however, is one of the most pressing ethical issues.

The "pastness" of food ethics might seem a problem. Food ethics, as it appears in the consumers' choice and actions, contains a systematic sluggishness, an inner inertness, as the damage must already have happened before the consumers can avoid or boycott it. In a way, by then it is too late. Food ethics points towards past time and makes up an "ethics of the past", an impossible ethic from the consumers' point of view, since what is already done cannot be undone: the food has been produced – under either ethically acceptable or unacceptable conditions. When the food ethics of consumers are based on sanctions, i.e. a sanction of specific products whose production history is found unacceptable rather than on the specific choice of products whose production history is known and appreciated, there is a danger that food ethics become the "least poor ethics". The choice is made among the products offered on the shelves of the stores, and all that can be done is to choose the least poor production history. Food ethics based on sanctions of productions that have already taken place is not the ethics of oneself but can only be the least poor ethics among the choices offered by the stores. Not to mention that in a globalized world, sanctions seem to have lost much of their effects: only in very few cases it has been possible to mobilize consumer boycotts that have been really efficient. For a sanction to be efficient there must be a linkage between involved partners and there must be something to sanction. Where there is no such relationship, sanctions become useless.

It is surely a fundamental characteristic of food ethics that it relates to production stories of lost time, but even so I consider this view of the consumers' possibilities too pessimistic. It might be a true and fundamental characteristic of the market that consumers only act when it is too late, but then one of the basic aims of food ethics should be to make it good again. Not in the sense that what is done can be changed and time rolled back, but that in the future, food production practices should be changed for the better. This points food ethics into the future as the vision of the good life with, and within, food production. In his book *Food Ethics* (1996), Ben Mepham, in an attempt to give a systematic overview of food-related ethical issues, sets up an "ethical matrix" shown in Table 2.

The principles mentioned (well-being, autonomy and justice) are indeed relevant issues and represent one of the first attempts to reflect systematically on food ethics in a principled approach, also sometimes referred to as *principalism*. For the moment, though, I am going to leave them uncommented on and instead concentrate on the question of the relevant actors within food ethics: the appropriate organisms, the producers, the consumers and living nature. Of these four, only three – the appropriate organisms, the producers and living nature – are directly involved in the production history.

The consumers are not usually involved in the food production. This is an important distinction within food ethics. Food ethics of course should be, and

Table 2. Ethical matrix showing three ethical principles and appropriate organisms (e.g. animals and crops), producers (e.g. farmers and employees in the food industry), consumers and biota[42]

	Well-being	*Autonomy*	*Justice*
Treated organism	e.g. animal welfare	e.g. behavioural freedom	Respect for *telos*
Producers (e.g. farmers)	Adequate income and working conditions	Freedom to adopt or not adopt	Fair treatment in trade and law
Consumers	Availability of safe food, acceptability	Respect for consumer choice (labelling)	Universal affordability of food
Biota	Conservation of the biota	Maintenance of biodiversity	Sustainability of biotic populations

are, concerned with the just distribution of food. However, the concern for nature, the living organisms and human beings involved in the food production is of another category. Precisely because they are the concern for those involved in the production, the actions or the shopping of the consumers are of immense importance – which is much less the case when it comes to fair distribution of food. I have already mentioned that in this book I have taken the view of consumers in order to see what possibilities are available for them to enact their food ethics.

However, as can be seen from the matrix the relevant ethical issues concerning consumers are availability of safe food, free choice of food and the affordability of food. These can be seen as rights of the consumers that put an obligation on the producers and retailers in the form of safe production systems, the duty to inform the consumers and so forth. The free choice of consumers (or as I would rather say the *informed choice* and the respect for consumer autonomy) is at the core of consumer food ethics because without information consumers cannot make any choice. The principle of consumer autonomy is the only above-mentioned aspect of the *ethics of consumption*, as it is sometimes referred to, that I shall deal with in this book.[43] Informed food choice points to the moral duties of the producers and retailers to give infor-

[42] Mepham, Ben: "Ethical Analysis and Food Biotechnologies: An Evaluative Framework". in Mepham, Ben (ed.): *Food Ethics*, p. 106.

[43] This means leaving out the other aspects of *consumer ethics* (e.g. food security and affordability), which are the ethics *for* consumers and the rights *of* consumers. See, e.g. Crocker, David and Toby Linden (eds.): *Ethics of Consumption*.

mation to the consumers concerning the production history and its food-related ethical issues.

As most people do not know anything or only have very sparse information about the production history, it is relevant to ask what then happens (which is the case of the food consumption). The answer is straightforward and well known to most people: food should be cheap!

Statistics show that every year consumers are spending less money on food.[44] They devote a smaller and smaller part of their income to food. For households seeking to maximize utility, this seems to be reasonable and justifiable: let us save money on food in order to be able to spend more on other things. The hidden assumption behind this is, of course, that food is more or less the same stuff anyway and that there is really no difference in quality. Moreover, it can indeed be difficult for modern consumers to see the difference between the foodstuffs. Let me try to follow the so-called utility maximizing consumers and their households to the outskirts, since this will indicate a number of things about the dilemmas facing modern consumers. For they are in a situation where they constantly have the feeling of being cheated. On the other hand, a lowering of prices would also lead to many other problems. If prices are lowered, the farmers and the food industry will have to produce more and will simultaneously have to lower their expenses. For their part, they have already gone a long way in this direction and their realignment has affected the welfare of animals, the well-being of the farmers and the job satisfaction of employees working in the food industry, the protection of the environment and the landscape and, of course, the quality of the food. Consumers either pay too much and have the feeling of being cheated or they pay too little and feel they have obtained inferior quality. The thought of being cheated torments consumers who, for this reason, prefer to swallow their pride and buy what is cheapest although they know it is inferior. But this at least saves them from the feeling of being cheated. It is not easy to be a consumer when price is the most important thing and the cheapest commodity is the worst.

The fitting up of the shops is tasteless, since the shopkeepers simply cannot afford to satisfy the fine taste of such cheese-paring consumers. The consumers lose their good appetite when shopping. Whatever the cost, it has to be cheap; this is why the consumers are shopping at the cheapest markets, well knowing that they naturally become dissatisfied. But they want the cheapest and therefore they arrive in large crowds to procure as much cheap food as possible of the poorest quality.

Because of the low profit yield on the sale of cheap food, the shopkeepers cannot pay the employees a decent wage and consequently feel constrained to

[44] See, e.g. Warde, Alan: *Consumption, Food and Taste. Culinary Antinomies and Commodity Culture*, p. 101.

buy the cheapest food. Even when they can afford something else, they know all too well that the profit margin on specialty items, delicacies and deluxe-quality foods is high. The quantity of products sold of this type is low; this condition makes the cost of distributing these items higher than the actual selling prices of the products. It is not easy to be a shopkeeper when one always has to please the consumers who come in looking for the cheapest products. They have to push the prices of the wholesalers down in order to put cheap food on the shelves of the food store.

This makes it difficult in turn to be a wholesaler. The wholesalers must import food that is produced in countries that are far away. This cuts costs, because consideration and care for the environment, the employees and the animals is virtually non-existent in these countries. It is hard to discontinue contracts with the local farmers, but one has to do so when it is imperative that the food be sold at those prices. In order to make the whole enterprise rational and profitable, the wholesaler is forced to take home larger portions than can be produced locally. The wholesalers know that the transportation of food all around the world causes pollution, but that is how the market works and we cannot escape globalization.

Nor is it easy to be a farmer when the wholesaler forces the prices down saying that he can get the goods even cheaper elsewhere, whether from the neighbour or from abroad. The farmers shake their heads at the prices in the stores when they consider what they themselves earn on the production of the food. In addition, it is hard to industrialize the small pigs and the chickens, but one must do so to stay in business. In order to make the production rational and efficient, farmers have to expand and go into big business and specialized production. They have just got to buy up the neighbour's family farm, but it is going to be very expensive, because the neighbour knows all too well that one is forced to expand, just to stay in business.

If there were anybody along this line who could agree to boycott the consumers and stop producing cheap pigs, it might be realistic to believe in a better food quality for the future. However, most of the producers and merchants have a hard time standing together; solidarity among competitors on the market is not very widespread. Moreover, globalization renders boycott old-fashioned and ridiculous. This describes many of the dilemmas pertinent to the modern food market and food production practices. It is a vicious circle, where selfishness, bookkeeping and economic rationality are assigned a higher priority than ethics. From the consumers' perspective the structures of the modern food markets keep the production history hidden. In their present state the structures of the market are unable to deal with the production history. Every time a food product changes hands almost all information of the former owner is lost. As things are today, it is difficult to make the production history visible; the effort needed to find relevant information concerning production practices is far too

Herculean a task for consumers. The complexity and the lack of transparency of today's food market make consumers distrust the producers and retailers. As consumers of food they only experience the food in its aesthetic and civilized form and never in an ethical perspective. It is no longer obvious that food has an origin and a history. To most consumers agriculture, the food industry and the other enterprises that are involved in bringing food to the food stores and the dinner tables are pretty much a closed book that they know very little about. They are unable to review the production history and thus equally unable to see into the future of how their food consumption is going to influence nature and society. The relations between producers and consumers are non-existent, and the coherence between the elements in the whole chain from "farm to fork" is concealed from the latter.

Food is not only a subject for the selfish consumer and household; it is rather an ethical subject with immense importance for society, nature and human beings. Food makes a difference; it is not indifferent in the sense that one kind of food is just as good as another. There is a difference in foods even though the food industry tries its best to standardize food.

Let me illustrate this with an example. Even such a very basic and relatively non-processed product as milk – at least most people do not think of milk as highly processed – is standardized for variations in breed, fat content, season, origin, feed, time of lactation, milking system and storage system, to mention but a few of the parameters that could otherwise make milk a very varied product.[45] The temporal variations in the level of proteins in the raw milk are standardized in the dairy to the same level. This enables the dairy to deliver the same homogeneous product all through the year and in all parts of the world. Milk is no longer a natural product but an industrial product; it possesses the most characteristic feature of mass-produced products: uniformity. The standardization of food products makes us forget their origin: food comes from living creatures and from living nature.

So why is that it is so difficult for industrialized and modern societies to deal with food ethics? From the producers' point of view the problem is that even the most ethically reasonable and justifiable food production practices will never be able to survive a financial deficit for longer periods. For this reason the ethics of the producers and the ethics of the consumers cannot be separated. The ethics of the producers cannot be maintained unless they are supported by the ethical consumption of the consumers. The ethics of the producers must therefore fit those of the consumers, otherwise there will be no customers. Some kind of conformity in the ethical standpoints of the producers and

[45] Sørensen, John (Arla Foods): "Milk – quality aspects relating to the raw material". Paper from the conference, *Kvalitet fra jord til bord (Quality from Farm to Fork)*, The Royal Veterinary and Agricultural University of Copenhagen, January, 2001.

consumers is necessary if food ethics are to exert their influence in the face of the economic rationality of the market. However, as production and consumption have been separated, the communication between producers and consumers is very limited and thus the chances of reaching agreements on food production practices are very poor.

PART II

THE INTELLECTUALIZATION OF FOOD

CHAPTER TWO

FOOD TO SCIENCE: ON THE INTELLECTUALIZATION OF FOOD

Most debates and public disputes about food tend to be diametrical. On the one side there is a materialist, positivist understanding of food and the specific understanding of rationality within science. This represents an intellectual relation to food, based on science. On the other side we find phenomenological and cultural understandings of food that appeal to the use of our senses, instincts, bodies, feelings, social awareness and so on.

The two different understandings of food have many parallels in the classical mind–body dualism, which is almost as old as philosophy itself and which acquired renewed significance in the controversy between the rationality of the Enlightenment in comparison with the sensitivity of Romanticism.

Prior to this, the philosophy of René Descartes (1596–1650) contained a dualism, which similarly separates the consciousness of mind and body:

> And although perhaps (or, as I shall soon say, assuredly) I have a body that is very closely joined to me, nevertheless, because on the one hand I have a clear and distinct idea of myself, insofar as I am merely a thinking thing and not an extended thing, and because on the other hand I have a distinct idea of a body, insofar as it is merely an extended thing and not a thinking thing, it is certain that I am really distinct from my body, and can exist without it.[46]

Even though Descartes considered the mind and the body to be closely linked, to the extent that he later asserted that they "constitute one single thing", this is only the case in a certain sense, because pain, thirst, hunger and other sensations perceived by the body "are nothing but certain confused modes of thinking arising from the union, as it were, the commingling of the mind with the body".[47] So even though the body may be very closely linked to the mind, more important for Descartes is the fact that thinking is separate from the body and can exist without it. Intellectual abilities and reasoning are considered part of the mind, whereas sensuality belongs to the corporal sphere, in which Descartes had no faith.

[46] Descartes, René: *Meditations on First Philosophy*: "Meditation Six: Concerning the Existence of Material Things, and the Real Distinction between Mind and Body", p. 97.

[47] Ibid. p. 99.

Mind and body constitute a dualism that can be found in many areas of human existence. Our relationship to food also tends to be characterized by this dualism: on the one hand the cool reasoning and calculating that can be used to maximize agricultural yields and maintain high levels of food security, and on the other hand the sensual, corporal and cultural aspects of food that are vital for the consumers' experience of closeness, coherence and comprehensibility. The scientific approach and the rationalization of agriculture do not leave much space for sensuality, closeness and coherence. And yet focusing too much on these qualities in food production and consumption might also marginalize rational and efficient food production practices, and in the worst case lead to an insufficient food supply.

"Intelligence" derives from the Latin *intelligere*, meaning to understand or grasp. Intelligence functions as an organizer of knowledge. The French phenomenologist Maurice Merleau-Ponty (1908–1961) states that objective thought is formed by removing the object from the experience and its appearance as an idea. The idea, i.e. common sense and the generalization of science,

> causes us to lose contact with perceptual experience, of which it is nevertheless the outcome and the natural sequel.[48]

There is a fundamental conflict between knowledge and sensuality. Thought is general and not subject to time, whereas sensuality is subjective and finite. We often experience knowledge and sensuality as exclusive opposites, so that when we make room for thought, sensuality and other outside impressions are excluded, as though they would hinder thought. Reflection must be excluded in order to make room for *full* corporal sensuality and physicality. Too many thoughts disturb the corporal presence in a given situation.

Sense perception and thinking are often presented as two different forms of cognition that have little in common. Sensation is frequently described as preconceptual, prerational and instinctive. It takes place *before* thought. Indeed it is common to say that the body *ap*prehends before thought *com*prehends. Belief in the ability of the intellect to achieve true cognition and knowledge has often thrown suspicion on the senses and physicality. Plato's philosophy is a classic example of this, as in *The Republic*:

> For I, for my part, am unable to suppose that any other study turns the soul's gaze upward than that which deals with being and the invisible. But if anyone tries to learn about the things of sense, whether gaping up or blinking down, I would never say that he really learns – for nothing of the kind admits of true knowledge – nor would I say that his soul looks up, but down.[49]

Contrary to the general view, a mistrust of the senses is also partially true of the natural sciences. Although their goal is to describe empirical reality, its ideal is not sensual. Sensuality is of course a prerequisite for empirical science, for

[48] Merleau-Ponty, Maurice: *Phenomenology of Perception*, p. 71.
[49] Plato: *The Republic*, 529B.

otherwise all access to the senses would be denied, but sensuality must be minimalized or neutralized, for it does not reproduce the world as it is but only the world as it is *for us*. Sensuality renders merely the appearance of reality, a manifestation – *die scheinbare Welt*. Sensual cognition is only a definition of the appearance of things. In contemporary natural science sensuality is replaced by representations, i.e. by measurements the result of which is mediated instrumentally as a value on a scale. The senses are only to be used to take readings from these instruments. At the same time the instruments may be regarded as an extension of the senses to the microscopic and macroscopic, to a level that man cannot otherwise attain via his senses; they are moreover a detachment from sensuality, so that the non-sensual is mediated via a visual representation. In natural science sensuality as subjectivization is unethical. As we shall see later, this attitude has significance for our relation to food and foodstuffs, and its influence is at work in agriculture, nutritional science, gastronomy and the consumption of food.

Both the intellectual and the phenomenological approaches to food are justified, I believe, and though they are fundamentally different, they should not be separated and placed in opposition to one another more than is necessary. The borderline between mind and body is not so sharply defined; intellectual reflection on food, for instance, is important for the phenomenological and narrative approaches that I shall develop in Chapters Four and Five.

First, however, let us proceed to a critical analysis of the intellectualization of food (there is no place for an anti-intellectual view!). The criticism below is a demonstration of how the intellectualization of food has focused mostly on scientific issues and therefore runs the risk of excluding other understandings of food, including cultural and phenomenological ones.

If we wish to understand the general critique of the technological development both of food (such as "novel food" or "functional food") and of food production practices as such, it is tempting to see it as a sign of how the cultural, ethical and sensuous relation to food has been marginalized in favour of a one-sided intellectualized and rational one. This latter is practised through the scientific mapping and analysis of chemical substances in food, through agro-science and industrialized food-processing practices, through the application of economic rationality, the widespread use of food declarations and so on. The charge is that food is being intellectualized on the premises of science and economics.

The study of nature, viewed as a systematic examination of nature, has exerted enormous influence on food itself as well as on modern relationships to food. Food *comes* from nature, it is a natural product, and thus in the philosophy of science the modern relationship to food follows in many respects the development within the studies of nature and life. If we are to undertake an investigation of how the discourse on food is dominated by modern scientific approaches, we should therefore begin by looking at the general scientific approach to nature and life.

When examining how the study of nature has been carried out down the ages, one is tempted to write a "history of biology". Soon however, one discovers that there is no science by the name of biology before 1800. It is perhaps understandable that the concept of biology does not exist in the ancient world. From the Renaissance to the end of the 17th century there is no biology either, only *natural history*. To call the study of nature from Aristotle and Theophrast down through the ages "biology" would be misleading. It would mean ignoring important steps in the development of man's relationship to nature.

This naming of biology was the result of new understandings of life (Greek: *bios*) and its unique characteristics. Biology designates the understanding of what separates living from non-living matter, based on the specific characteristics of the living cells, the different forms of reproduction, metabolism, the working together and functioning of different organs in the body and different organisms in nature, etc., concepts of life that were gradually put forward during the second half of the 18th century. The study of animals and plants in antiquity and the natural history of the Renaissance and the Enlightenment were indeed concerned with the living, but had not yet become biology, the science of life, since in these periods there was no clear distinction between animate and inanimate.

The history of the study of nature can be told in many different ways, for there are various aspects to it. Bearing in mind the overall theme of this chapter, the intellectualization of food, I will focus on the development in methods, theories and perceptions of the living within the study of nature.

1. THE HERMENEUTIC APPROACH OF EARLY NATURAL HISTORY

The application of the concept of *history* in connection with the study of living beings is first found in Aristotle's *History of Animals*. Aristotle's zoological works were extensive and listed around 550 different animal species. He was generally less successful with plants, but with animals he demonstrated such knowledge that Charles Darwin is once claimed to have said of the two great naturalists, the Swede Carl von Linné (1707–1778) and the Frenchman Georges Cuvier (1769–1832), that they were mere schoolboys compared to Aristotle.

Sensuality occupied an important position in Aristotle's descriptions of nature. Of the five senses in man he regarded touch as the most precise, followed by taste. But all the senses could be used in a description of the nature of the matter in question: rawness, smoothness, softness, hardness, weight, sound, taste, form, etc. According to Aristotle all flora and fauna are regarded as a combination of matter and form. The matter is described via a combination of the four elements: earth, water, fire and air. The elements are linked through the perceptions of wet, dry, hot and cold: the earth is cold and dry; water is cold and wet; air is hot and wet; and fire is hot and dry.[50]

The History of Animals and other books by Aristotle about fauna are not just classifications of animals but also descriptions of their lives, their interaction with their surroundings and their significance for man. The classification is built on their external and internal characteristics, on their surface morphology and their inner anatomy, on their sensory organs, their sounds, their sleep (including their hibernation), their reproduction, their movements and habitats, their food, their sicknesses, their characteristic features and so on. The animals' physiognomy is often described in relation to their physical abilities and way of life: "Of insects, such as have teeth are omnivorous; such as have a tongue feed on liquids only".[51] The appearance of the creatures is also related to their mental character: "Soft hair indicates cowardice, and coarse hair courage".[52] Aristotle assumes that there is accordance between their appearance and their mental abilities, and for him there is nothing strange in relating their appearance to their characteristics, virtues and ethics.

Aristotle's concept of history as it appears in *The History of Animals* belongs to its age; he did not record events over time, or any development of the species. Antiquity was mostly interested not in changes and events but in the unchangeable and the eternal. *The History of Animals* is not an evolutionary story; indeed, it is not even history in the present sense, but a recording of observations and a classification of animals. The ancient Greeks understood history first and foremost as watching and observing, and then noting down these observations, more a kind of natural phenomenology than a history. Their conception of history was to see and narrate what was seen, a sort of phenomenological observation of things as they are. From antiquity and right up to the end of the 18th century the concept of history differed radically from today, where it is most often regarded as a reconstruction of a sequence of events.

Certainly, people in the past were capable of telling stories; indeed storytelling and (hi)stories of the past were *narrative*:

> [T]his was the function of history for about two thousand years, a school in which one could become prudent without making mistakes.[53]

According to the German historian, Reinhart Koselleck (b. 1923), historiography as a scientific discipline only came into being around 1780. Koselleck links the change in the meaning of the concept of *history* to the desire to predict, a trait that emerges in the Age of Enlightenment. In this period man would no longer subordinate himself to his fate; he realized that the future would not be like the past, and that through his own inventions and with the aid of science and technology he

[50] Aristotle: *On Generation and Corruption*, 330a30–330b5.
[51] Aristotle: *The History of Animals*, 596b12–13.
[52] Aristotle: *Physiognomonics*, 806b1.
[53] Koselleck, Reinhart: *Futures Past. On the Semantics of Historical Time*, p. 22.

could become the creator of his own history. Prediction of future events that could be the result of actions in the present was important in order to make reasoned, rational and emancipated decisions. The ability to predict and control future events became essential with the emergence of natural science and technology.[54]

Also the concept of "natural history" in the late Middle Ages and the Renaissance was still employed as in antiquity. Natural history was the description *of* nature and certainly not the intervention *in* nature. Nor did the Renaissance add much more to the knowledge of the individual than that the four elements had a special relationship to each other that was characteristic for each species – though precisely how remained unclear. In many ways the Renaissance scholars merely repeated what Aristotle had already noted. Matter in itself could not be used to describe and classify animals and plants, for as long as man was limited by the perception field of his sensory organs, it was impossible to explain the various types of matter. It was thus the *form*, or as modern biology terms it, the *phenotype*, and not the substance, that determined the description of flora and fauna,[55] *a*nd morphology that formed the basis of the *visible* structures of living beings.

Nevertheless, there was *one* new development in natural history during the Renaissance.[56] *F*or although the visible structures were still regarded as surface phenomena in the form of signatures and signs that could form the basis of a classification, behind these visible structures and forms there was *another* relation between things, a coherence that did not reveal itself directly or immediately to man. According to Michel Foucault (1926–1984), the Renaissance mind believed that the surface of things bore the sign of what was most essential in them. To go behind things and find their "true nature" one could therefore use their form, their signatures, not as they appeared in themselves, separated from the rest, but as signatures in relation to one another, i.e. in a coherence. In Foucault's analysis in *The Order of Things* (1966) this relation between signatures reveals itself at four different levels, namely:

(1) Convenience
(2) Emulation
(3) Analogy
(4) Sympathy

Convenience was seen in the Renaissance as a link between things in nature; it described how things came closer to, and bordered on, one another. Nature

[54] Koselleck, Reinhart: *Futures Past. On the Semantics of Historical Time*, 198–203. Koselleck's analysis is mainly restricted to German literature and thus to the genealogy of the German word *Geschichte*.

[55] On the other hand the use in modern classification of genetic material is said to be a classification by substance or by genotype.

[56] This presentation builds on Michel Foucault's *The Order of Things* (especially ch. 2) and François Jacob's *The Logic of Life* (especially ch. 1).

was organized according to a continuity principle based on Aristotle's Scala Naturae; in the Renaissance all transitions were gradual, never sudden. The external signifiers, the forms of things, showed how animals and plants were linked to each other. This must not be seen as the beginnings of a concept of evolution, for even though transitions were gradual, the species had been created as they were *for all time*. The continuity principle brought together in the classification those things that resembled each other. In describing plants, animals and people, the similarities were stressed as much as the differences; and occasionally the former was even emphasized at the cost of the latter. Everything was nature and nature was one and coherent. Hieronimo Cardano (1501–1576) explained how parts of a plant corresponded to parts of an animal, the leaves of the tree to the hair of animals, its roots to the animal's mouth, etc.[57] The plant was almost like an animal, only standing on its head. And as for the animal, it resembled man; and man's approach to the universe, as Giambattista della Porta (1535–1615) wrote, was as follows:

> As with respect to its vegetation the plant stands convenient to the brute beast, so through feeling does the brutish animal to man, who is conformable to the rest of the stars by his intelligence.[58]

This leads us to the idea of the *emulation* of things by one another. The emulation was not merely passive; it was a copying or a mimesis. The animals were emulated in the stars, the planets and in the very stones. All created beings emulated one another.

Analogy in the signatures demonstrated the kinship between all visible things. There were plants that resembled snakes and eyes. These were visible signs that, if man could interpret them, would help him to understand the hidden coherences and powers of nature. Apparently invisible coherences were visible to whoever understood how to interpret nature. The kinship between aconite and eyes would have remained unnoticed had it not been for the fact that the seeds of the aconite resemble human eyes. The reason for these resemblances was that they contained the same attributes; the aconite was thought to heal eye diseases.

Sympathy and *antipathy* were forces that respectively united and separated things. Sympathy was an attraction, as when flowers turned to the sun or roots bored for water. But sympathy would make things the same and erase the differences between them if it were not kept in check by antipathy. Antipathy maintained the specific characteristics of things and kept them separate from one another. Antipathy was the struggle between things, such as the tree's struggle for light, space and water.

[57] Jacob, François: *The Logic of Life*, p. 23.
[58] Porta, Giambattista della: *Magie naturelle* (1650). Quoted from Foucault, Michel: *The Order of Things*, p. 19.

In the Renaissance, nature was to be interpreted rather than explored. Natural history was the exegesis of the secret signs of nature, a sort of natural hermeneutics. A living being represented a hidden world, with ramifications to the rest of the world, to other flora and fauna, to the stars, to the world of man, which should be made plain via interpretation. When the systematic zoologist Ulisse Aldrovande (1522–1605) described the horse, he thus spent only 4 pages on describing its form, but nearly 300 in order

> to relate in detail the horse's names, its breeding, habitat, temperament, docility, memory, affection, gratitude, fidelity generosity, ardour for victory, speed, agility, prolific power, sympathies, diseases and their treatments: after that the monstrous horses appear, the prodigious horses, fabulous horses, celebrated horses, with descriptions of the places where they won glory, the role of horses in equitation, harness, war, hunting games, farming, processions, the importance of the horse in history, mythology, literature, proverbs, painting, sculpture, medals, escutcheons.[59]

The history of a plant or an animal was not only a description of its observable phenomena but just as much a description of everything that had been seen and heard about it, and all that had been told about it through nature and man. Acoording to Foucault: "[N]ature, in itself, is an unbroken tissue of words and signs, of accounts and characters, of discourse and forms".[60] At this time it was unnecessary to distinguish between knowledge acquired in various ways.

For Aldrovandi and others, such as Konrad Gesner and his *Historia animalium* (1551–1558) or Pierre Belon and his *Historie de la nature des Oiseaux* (1555), natural history comprised the entire interwoven network constituted by living beings: their morphology, organs, medical application, their place on any escutcheon, their spiritual nature, their myths and legends and much more. For Foucault this was their definition:

> The history of a living being was that being itself, within the whole semantic network that connected it to the world.[61]

No distinction was drawn between observation, document and fable, because the sign had not yet been separated from the thing itself, had not yet become what we today know as a representation of something else. The sign was in the thing and the thing was in the matter, the sign and the thing were the same. The word *horse* was not separate from the animal, and legends about the horse were part of the horse itself. The curative powers of the aconite on the eye were part of the plant itself, confirming the convenience between the seed of the plant and the eye.

How can we understand the Renaissance view of nature today? Is it so foreign as to be impossible to translate into the language of our time? Foucault

[59] Jacob, François: *The Logic of Life*, p. 22.
[60] Foucault, Michel: *The Order of Things*, pp. 39–40.
[61] Ibid. p. 129.

attributes the difference between the sign *being* the thing and the sign *representing* the thing to the arrival of modern natural science and of modernity itself. According to *The Order of Things* this is history's most radical revolution of cognition, with consequences that we have not yet completely understood. Knowledge acquired a new meaning and a new content. Through representations the signs were dissolved from the things, from the concrete. The pre- and post-Renaissance differences can be depicted as follows:

The sign is the thing: Lack of distance, corporality, concreteness, sensuality, the emotional element, coherence and totality.

The sign is a representation: Reflection, self-awareness, the intellect, empirical rationality, distancing and the detached.

So natural history was not history writing but hermeneutics with a certain content of phenomenology. In the Renaissance the phenomenological description of nature lost ground to this speculative hermeneutics – or perhaps we should call it, this imaginative biosemiotics. Living beings were created once and for all, fixed in their forms and with no development. Of course time existed, but living beings did not change and therefore had no history. New generations were repetitions of previous ones. This was a general observation that everyone could make. The sparrows and the beech trees had the same appearance as in grandfather's time. Variations might be observed, but the essence of the species was unchanging through the generations.

François Jacob (b. 1920) describes this idea of natural history regarding the development of animates in very categorical terms:

> It is no exaggeration to say that until the eighteenth century living organisms had no history.[62]

It is not until Lamarck (1744–1829) and others in the 19th century that the idea arises of a gradual *alteration* in the development of the species – in the form of Lamarck's ideas of inherited characteristics. And it is not until Darwin (1809–1882) and Wallace (1823–1913) in the 1850s that the idea of a continuity between the species is presented clearly and unambiguously.

Jacob writes that the modern conception of the history of animates and temporal beings is linked to four things: origin, continuity, instability and chance.[63] Origin, because life does not come from nothing, but from something – and at a point in time. Continuity, because life does not continually come into being,

[62] Jacob, François: *The Logic of Life*, p. 131. It was not until the mid-18th century that the first descriptions of nature appeared which suggested that natural historians of the time perhaps had a vague feeling that animates had an evolutionary history. Through the study of fossils geology was giving birth to new theories on the disappearance of species due to floods and other natural disasters.

[63] Ibid. p. 130.

but already existing life reproduces itself and lives on. Instability, because animates do not just copy themselves, but now and again create new variations that are necessary for any development. Chance, because the development of animates cannot be presumed to have a specific intention, but follows an unpredictable and unplanned pattern. In contrast to natural history, the history of animates, i.e. evolution, becomes nature's own history, the inscribing of the species in the great perspective over millions of years.

2. THE PHENOMENOLOGICAL APPROACH OF LATE NATURAL HISTORY

However, natural historians do not continue their mere exegesis and interpretation. During the Enlightenment natural historians slowly became more interested in understanding nature's own order, and so they set out to investigate the categories of nature by systematizing nature into natural classes. In this context categorization can exemplify how natural history developed from a hermeneutic to a phenomenological discipline. But let us start with classification in ancient times.

It goes without saying that there is nothing unique about naming and dividing nature's living creatures into categories; indeed it is a feature of all cultures. The criteria for classification among so-called aboriginal peoples are typically based on the practical value of the various genera and species as nourishment or medicine. The Argentinian writer Jorge Luis Borges (1899–1986) refers to an old and probably fictive Chinese encyclopaedia, *Celestial Emporium of Benevolent Knowledge*, which contains a rather unusual classification of animals:

(a) belonging to the emperor
(b) embalmed
(c) tame
(d) sucking pigs
(e) sirens
(f) fabulous
(g) stray dogs
(h) included in the present classification
(i) frenzied
(j) innumerable
(k) drawn with a very fine camel-hair brush
(l) et cetera
(m) having just broken the water pitcher
(n) that from a long way off look like flies[64]

[64] Borges, Jorge Luis: *Other Inquisitions*.

As well as being amused by the absurdities here, the reader is admitted to another universe with an astonishing and radically different experience of the world. At the same time there is room for a wry smile at the way our culture typically regards the classification systems of other cultures as being illogical.

As we have seen, it is Aristotle who produces the first known research and classification of nature's living beings in a form that we would today call "scientific". His work left a long-standing mark, with his classification of animals remaining largely unaltered for over 2000 years.[65] This classification of nature's living beings is seen as an organizing project based on sense experiences and instincts:

> We must attempt to recognise the natural groups, following the indications afforded by the instincts of mankind, which led them for instance to form the class of Birds and the class of Fishes, each of which combines a multitude of differentiae, and is not defined by a single one as in dichotomy.[66]

In Aristotle's time a much-discussed subject was the classification of nature's living beings into oppositional divisions, such as *flying* as opposed to *non-flying*. The dichotomy was employed in logic and in various other areas, and it was therefore reasonable to use it on nature too. But Aristotle rejected this approach in the case of flora and fauna, because in his view living beings were not created according to this form of logic. Instead of oppositions Aristotle stated that "it is generally similarity in the shape of particular organs, or of the whole body, that has determined the formation of the larger groups".[67] Aristotle's attempt to acquire knowledge of plants and animals led him to establish a "natural" order based on phenomenological differences and similarities.

Even though we do not find anywhere in Aristotle a systematized table of his hierarchical division of nature, posterity termed his classification in the mentioned *Scala Naturae* and presented it in large tables. *Scala Naturae* also included the inanimate because

> there is a mean between life and inanimate, because the inanimate is that which has no soul nor any portion of it. But a plant is not one of those things which entirely lack a soul, because there is some portion of soul in it; and it is not an animal, because there is no sensation in it, and things pass one by one gradually from life into non-life.[68]

In Aristotle's thought it was the soul that was the difference between animate and inanimate, a distinction that we still consider reasonable, though without justifying it by referring to the soul.

However, Aristotle's division more or less disappeared during the discussions on scientific theory in the Enlightenment. Whether it was an animal,

[65] Maule, Anne Fox and Peter Wagner: "Renæssance til revolution", in *Naturens historie fortællere*, Vol. 1, p. 129.

[66] Aristotle: *Parts of Animals*, 643b 10–14.

[67] Ibid. 644b 8–9.

[68] Aristotle: *On Plants*, 816b1.

plant, mineral or machine, in the Cartesian view all bodies were made of a substance, and he therefore believed that it was the same laws and rules for movement that controlled both animate and inanimate; it was thus impossible for him to distinguish between animate and inanimate, life and lifeless. This position influenced the entire period up to 1800.

In the 18th century there were two main scientific directions that were concerned with living beings: physiology, which had its source in medicine and which studied the composition of bodies; and natural history, which described and systematized nature on the basis of simple visual observations. Natural history had the best conditions and developed faster than physiology, among other reasons because of the favouring of the visual above all the other senses.[69] Precisely because the focus was on the visual appearance of things, the differences between animate and inanimate were unclear; a sharp distinction between the two was not on the cards. There were so many other ways in which to divide up the world that seemed equally relevant. Natural history was gradually detached from the similarities, conveniences, emulations, analogies and sympathies or antipathies characteristic of the Renaissance, so that nature could appear naked before the human eye. Stories, proverbs, myths and their uses were no longer accorded the same status as *observation*.

What counted in the late (or second) period of natural history was the visible structure and classification of things and that alone. The understanding of what was legitimate knowledge was redefined. Of course nature must be decoded, but not with the Renaissance focus on similarities and early forms of speculative and imaginative natural semiotics, which post-Renaissance natural historians regarded as sense illusions. Increased importance was attached to converting into figures the measurement, enumeration and organization of knowledge, a process that was slowly spreading into other sciences as well. In place of similarity was *difference*. Cognitive ability came to rest solely on the ability to distinguish, meaning in turn to analyse and reduce in order to *identify* (again in contrast to the uniformity of the Renaissance). Foucault puts it thus:

> [T]he written word ceases to be included among the signifiers and forms of truth; language is no longer one of the figurations of the world, or a signature stamped upon things since the beginning of time. The manifestation and sign of truth are to be found in evident and distinct perception. It is the task of words to translate that truth if they can; but they no longer have the right to be considered a mark of it.[70]

From the mid-17th to the mid-19th century there is a whole string of scholars contributing to natural history: Jan Jonston: *Historia naturalis de quadripedibus* (1657), John Ray: *Historia plantarum generalis* (1686), Georges-Louis

[69] Hans Jonas in *The Phenomenon of Life* (chs 6 and 7) describes how sight was the preferred and favoured sense ever since antiquity.

[70] Foucault, Michel: *The Order of Things*, p. 93.

Buffon: *Histoire des Animaux* (1748) and *Histoire naturelle* (1749), Michel Adanson: *Histoire naturelle de Sénégal* (1757), Jean-Baptiste Lamarck: *Mémoires de physique et d'histoire naturelle* (1797), Fabricius: *Resultate naturhistorische Vorlesungen* (1804), George Cuvier: *Histoire de poissons* (1828–1849). These natural historians worked primarily with phenomenological structures as the basis for the description of the animate, rejecting the comprehensive, interwoven and imaginative science of the Renaissance in favour of presenting the animate pure and simple, purged of irrelevant "stories", isolated and detached from other contexts and stripped to the human eye. Where Renaissance natural history had been largely narrative – not "history" but the story of how living beings interacted with the rest of the world – it now became non-narrative, and all ramifications and excursions to other areas of existence were first limited, then neglected and finally abandoned. The visible was studied, described precisely in words and figures, named and categorized and related to its external wholeness and its application.

We might wonder why the flora and fauna of the world as described by the French natural historian Georges-Louis Buffon (1707–1788) and the Swedish botanist Carl von Linné in his *Systema naturae* (1735) had not been described earlier, for their morphological features had been observable at any time. It cannot be because people were better able to see in the 17th and 18th centuries; it must be because sight was now trained on something new. The focus moved from the similarities and analogies of the Renaissance that aimed at revealing nature's invisible coherences to Linné's interest in the organs of reproduction, the androeceum of flowers and the gynoeceum of fruit leaves, in order to create an *artificial system* based on differences (not similarities). He wrote that the natural historian distinguishes

> "with the aid of sight between the parts of the natural body, describes them in number, figure, proportion and position and names them".[71]

Fables, second-hand sources, proverbs and their uses were placed last in Linné's method, in which every creature is described by name, theory, family, species, special characteristics, use and literature, the last two reflecting their low priority. But the other senses too were marginalized; smell, hearing, taste and touch were all secondary to sight: they may well have been used but only in individual cases and not in a systematized and theoretical way.

One of the important discussions of cognitive theory during the 18th century concerned the division of animates into species and families. Buffon doubted the reality of a continuous *Scala Naturae*, believing instead that in nature only the individual existed; concepts such as kinds of families, orders and classes are only found in our imagination and are, so to speak, artificial. This nominalist view says that only what can be experienced and understood through the senses in reality exists. Only what can be sensed belongs to the physical world.

[71] Linné, Carl von: *Systema naturae*, p. 215.

Again we see how the intimate coupling of things and signs in the Renaissance is replaced by the sign as a representation in the 17th and 18th centuries, and how the phenomenological-based taxonomy was regarded as a way of representing the order of nature.

3. BIOLOGY AND THE INVISIBLE CHARACTERISTICS OF LIFE

It is not absolutely clear as to when the concept of *biology*, the science of life, replaced natural history. By 1802, it was used by a number of natural historians, including Gottfried Treviranus (1776–1827) and Lorenz Oken (1779–1851), but it is most often credited to Buffon's pupil, Jean-Baptiste Lamarck (1744–1829),[72] whose declared goal was to find the laws that controlled "la marche de la nature". The appearance of the concept of *biology* marked the recognition that animate and inanimate could no longer be regarded collectively, that there was no gradual, continuous transition linking animate to inanimate. What led to this was the distinction made by a number of natural historians such as Lamarck, Charles Bonnet and Antoine Laurent de Jussieu, as well as the philosopher Immanuel Kant (1724–1804) in the second half of the 18th century. This distinction was based on differences in *organization*.[73] It was the distinction between the unorganized and the organized, the inorganic and the organic. In his *Philosophie zoologique* (1809) Lamarck writes:

> One will first remark a large number of bodies composed of raw, dead material which increases by the juxtaposition of the substances forming it and not because of any internal principle of development. These beings are generally called *inorganic* or *mineral beings* . . . . Other beings are provided with organs appropriate for different functions and are blessed with a very marked vital principle and the faculty of reproducing their like. They are comprised in the general denomination of *organic beings*.[74]

Lamarck's use of the concepts "inorganic beings" and "organic beings" shows how difficult it was, and how much it was still in its infancy, to distinguish between animate and inanimate; both minerals and animates were regarded as beings.

It is reflections such as these that make Foucault declare that "up to the end of the 18th century, in fact, life does not exist: only living beings".[75] Bearing in mind that as an authority on natural history right up to the 16th century Aristotle himself actually distinguished between animate and inanimate, this rhetorically subtle remark requires further clarification to see whether or not the propensity

[72] It is doubtful, however, whether it is Lamarck who is the first to use the term. According to Mittelstraß, Jürgen (ed.): *Enzyklopädie Philosophie und Wissenschaftstheorie*, Vol. 1, p. 315, it was also used by K.F. Burdach in his *Propädeutik zum Studium der gesammten Heilkunst* (1800).

[73] See Jacob, François: *The Logic of Life*, p. 100.

[74] Quoted in Jacob, François: *The Logic of Life*, pp. 86–87.

[75] Foucault, Michel: *The Order of Things*, p. 160.

towards the great beating of wings in the history of ideas has not been exaggerated here. For Aristotle it was the presence of the soul and sense perception that characterized life; they expressed nothing about matter itself but rather something *through* it. When Aristotle defined life, it was through specific attributes in matter and not matter itself (which always consisted of the four elements). For Aristotle it was not just living beings but life itself that was dissociated from the inanimate. In the flourishing of natural history from the mid-16th to the mid-18th century, there were a number of things that prevented life from dissociating itself into a special category, so that it was precisely the living beings and not *life* that were thematized. It would be too simple to claim that it was technically impossible to study the specific features of animates and thus life itself; the microscope had been invented in the mid-17th century and could magnify objects sufficiently to facilitate the observation of individual cells. But this practical invention did not immediately win recognition, and paradoxically his first observations by microscope led Gottfried Leibniz (1646–1716) to reject the idea of a distinction between the organic and the inorganic, as Descartes had already done on a more theoretical basis in the first half of the 17th century. Observations through the microscope were interpreted via the continuity principle of the gradual transition between life and non-life and hindered a distinction between the two. If there was no reason for a distinction in substance between life and non-life, one could imagine two ways of conceiving substance: One could follow Descartes in thinking that all matter was subject to the mechanical law that had already been observed in inanimate nature. The regularity of mechanics could, so to speak, be transferred to living beings. Or one could say that all things are animate, as the above quotation from Lamarck suggests; that all things are beings, whether they are organic or inorganic.

Although before 1800 nature was divided into animate and inanimate, the uniformity of matter established a coherent totality; as regards matter, nature was one. Man may differ from stone, but both belonged to the same nature, and there was no insistence on a sharp distinction between living and non-living nature, between animate and inanimate. They were two different categories, but they were interwoven with each other because nature was continuous; all links were gradual. Nature made no sudden leaps. Before the appearance of biology, nature was not only coherent but was also made of the same substance.

The establishment of biology as a scientific area of study, as the life sciences, *la science du vivant*, foreshadowed the separation of life from other substance, from the inanimate. It recognized that living beings and inanimates could not be described according to the same principles. It was a distinction between two essentially different forms of *materiality* based on differences in the organization to which matter was subjected. Of course, even before the 19th century, scientists were aware of the differences between man and stone, between a dead object and a living being, but the difference was not "substantial" and could not be explained with the aid of matter; even though there was a difference between man and stone, both by nature were made of the same four elements in different proportions.

How then did the distinction between animate and inanimate come about? What sort of reflections led to the definition of life as organization? The answer is that it was connected to a shift of focus in the study of living beings; their distinctive characteristics were no longer sought in their conspicuous surfaces, but in their depths. Scientists now moved into the interior of matter and the body. Matter was subjected to chemical analyses, first by the alchemists, and later by the chemists. The French chemist Antoine-Laurent Lavoisier (1743–1794) began his study and division of the chemical substances on the same principles by which plants had been divided. He defined the simplest substances with the simplest words,[76] an approach that is still valid. Chemistry gradually became established as a science capable of classifying, naming and measuring the characteristics of each substance. The German chemist Justus von Liebig (1803–1873) is often credited with turning organic chemistry into an independent discipline in the course of the 1830s.[77] Empirical analysis moved inwards into the organisms and studied the interaction between the organs, which gave the totality its meaning,[78] and where each separate part had its own *function*. The study of the visible structures shifted towards the *functional*, which became the invisible essence of animates. Where natural historians had previously classified animates, the new biologists set greater store by the knowledge of the organization of living organisms. The teleology of biology, its purported aim and its distinction between animate and inanimate was now functionality, which led to a consequent interest in *causal explanations*. It was the function of the heart to pump blood round the body, as the English doctor William Harvey (1578–1657) already discovered in 1628. The causal explanation was now found in the muscles of the heart, the supply of blood, the nerve paths and so on. The combination of functionality and agents, the purpose and casual explanation in the same organism, separated the life sciences from the other sciences such as physics, mathematics and chemistry.

With functionality the organisms became the bearers of their own purpose, their own intent. But this did not mark a return to the sign of hidden essences of the Renaissance, since it was now the *organization* that was hidden behind the visible structure of animates. If the scientist or biologist proceeded into the organism, as when the anatomist Georges Cuvier in his comparative dissections compared organs and their particular purposes, scientists could no longer settle for a purely phenomenological experience of the structure of things, for the functions of the organs were not apparent from their visual structure, from their form or other characteristics. Organization as a principle for animates meant at

[76] Lavoisier, Antoine Laurent: *Traité de chimie*.
[77] For example in *Anleitung zur Analyse organischer Körper* (1837).
[78] Jacob, François: *The Logic of Life*, p. 74.

the same time that biology did not gain the same relationship to phenomenology as natural history had had during the Enlightenment. Biological research no longer consisted primarily of describing the visible in order then to name it and add it to the hierarchy of taxonomy. The study of the inner functionality of animates heralded the end of natural history (which was not the writing of history) and the beginning of biology and nature's own history, evolution.

Within philosophy Kant too looked at the organization of animates, though from a somewhat different angle. In the second part of *The Critique of Judgement* from 1790, he discusses the functionality and expediency of nature. We can guess that he was inspired by the debate of the day on the inner organization and functionality of living beings, but one of the things that enables man to wonder at nature, he wrote, is that it does not have just one purpose for us; and not only does it present itself as useful to us but it also seems as if it has a purpose in itself. According to Kant, such a purpose, lying outside man's own interests, is to be found where both cause and effect are present in the same thing,[79] unlike external causality where they are separate. Cause and effect in the very same thing are to be found in living beings; they are not exclusively controlled by an external causality. Living beings *organize* themselves: each organ of a living organism conditions another and forms itself out of itself.[80]

This organization of a living being refers to the internal relations within the organism. Biology abandons the representations of the natural historians – the sign as a representation of order – because it no longer makes sense to study animates on the basis of their surface structure as long as life's specific character lies in its internal organization. In this way biology opens up for the organizational room of animates, first the organs and then the cells. But biology does not thereby abandon the dominance of sight; it merely directs it towards what is not immediately visible but which can be *made* visible with the aid of various techniques.

The separation of life from inanimate matter means that life may be defined by its opposite, death. The Parisian doctor Xavier Bichat (1771–1802) affirmed in his famous definition that life "is the sum of the functions that oppose death".[81] Life is defined by its struggle against its dissolution and destruction, the struggle against being consumed by the inorganic, the struggle against death. Life establishes an order that is apparent in its organization. When flora and fauna were studied during this period, it was in order to study life itself, to study how animates were separate from the dead, and biology from physics.[82] But to define life as organization and a contrast to death does not provide a sufficient explanation for life so long as one remains at the level merely of

[79] Kant: *The Critique of Judgement*, pp. 370–371.
[80] Ibid. pp. 373–374.
[81] Quoted by Jacob, François: *The Logic of Life*, p. 90.
[82] Jacob, François: *The Logic of Life*, p. 92.

organ and tissue. For this does not define the peculiarity of life. So from lack of knowledge a number of scientists preferred mechanical explanations.

Others attributed to life a specific life force, believing that the animate could be explained on the basis of the inanimate. Lamarck, for instance, operated with "subtle fluids" as an explanation of life. In 1908 the German biologist Hans Driesch (1867–1941) dubbed this type of explanation "vitalism". Driesch himself worked on the idea of a "form-creating principle" as characteristic of the animate. The mechanistic and vitalistic views often acted as antitheses in the 19th-century discussions on the definition of life. The vitalists pointed to the totality, the complexity and the creative unpredictability of the animate,[83] but vitalism did not possess the same power as mechanism to explain, reveal causal links, predict and manipulate. Vitalism contained a criticism of the tendency of mechanistic biology to regard life as mere matter. However, despite the poetic power of the vitalistic theories of the time concerning life and the life force, they must be said to be speculative and difficult to apply practically. When Aristotle writes that "the soul cannot be without a body, while it cannot *be* a body; it is not a body but something relative to the body",[84] we can say that the life force is neither free of the animate nor is itself animate, but is something *through* it. Vitalism and functionalist biology studied in reality two different sides of the animate: matter and something that exists through matter that is not matter itself. On this understanding it must be said that vitalism is not dead but has been continued and developed. Today there are a number of natural science disciplines linked to biology which can be said to study something connected (relative) to matter, but which is not matter itself: biosemiotics, artificial life, autopoisis, theories of emergence, bioinformatics and cybernetics. To regard deoxyribonucleic acid (DNA) as information or signs is in fact to claim that there is something about matter that is not matter itself but something else.

4. THE END OF PHENOMENOLOGY IN BIOLOGY

Natural history and biology deal with living beings that develop in space: they are made of organic matter and exist in an environment. The phenomenological is clearly apparent in Aristotle:

> [R]especting perishable plants and animals we have abundant information, living as we do in their midst, and ample data may be collected concerning all their various kinds, if only we are willing to take sufficient pains.[85]

Aristotle did not doubt the epistemological status of nature: that nature exists, it would be absurd to try to prove; for it is obvious that there are many things

[83] Cf. Bergson, Henri: *L'évolution créatrice*.
[84] Aristotle: *On the Soul*, 414a19–21.
[85] Aristotle: *Parts of Animals*, 644b27–31.

of this kind.[86] This is why Aristotle's natural history was sensuous and phenomenological, a description of the living beings' appearance in space. The phenomenological approach also meant that he was particularly focused on totalities and not on smaller detached parts.

In the Renaissance the space of nature was expanded, so that it was no longer enough merely to describe the appearance of living beings in space phenomenologically. Natural history in the Renaissance did not confine itself to the phenomenological; rather the phenomenological description became a very small part of natural history. Instead came the cultural links and meanings of living beings to occupy a place, such a significant place, one might add, that the importance of living beings in the culture is placed in the centre at the cost of the observation of nature itself. The description of living beings was secondary to the importance they had for people. Renaissance natural history described the *social space* that the living beings occupied.

The trials against animals in the late Middle Ages and the Renaissance show this very clearly. In 1546 and 1587 the inhabitants of Saint-Julien in France took legal action against a colony of weevils that had invaded the vineyards and caused considerable damage. The peasants wanted the lawful expulsion of the tiny beasts. These are just two of a number of trials that the French philosopher Luc Ferry (b.1951) relates. Most of the trials took place between the 13th and 18th centuries.[87] Ferry uses these thought-provoking stories in a critical analysis of the misanthropic attitude of the environmental movement; he describes the trials as prehumanistic, absurd and ridiculous. Nature cannot be made legally responsible; it has no stances. It is meaningless to speak of "the freedom of nature" or its voluntary intention. Ferry's critique does not leave much room for a historical understanding of the trials, i.e. in the light of how nature was perceived at that time. The perception of nature in the natural history of the late Renaissance gives us reason to believe that the trials were the results of an extremely *socialized nature*. Nature was socialized to the extent that living organisms could be made legally responsible. Renaissance man was not interested in nature itself as it appears to us from a purely phenomenological perspective; he was not interested in the autonomy and self-organization of nature that was later to be the perception of the science of biology. He was interested in the nature, which meant something to him, and finding the meaning hidden in nature. The role of nature was not that of a scientific object but was rather an integral part of the world.

The arrival of modern philosophy changed this radically. In his *Meditationes de Prima Philosophia* (*Meditations on First Philosophy*) (1641/42), Descartes exercised a methodical doubt about the possibilities of cognition, even the

[86] Aristotle: *Physics*, 193a2–4.
[87] Ferry, Luc: *The New Ecological Order*, pp. IX–XVI.

existence of physical objects and corporality, for the senses might be susceptible to the deception of a wicked demon. They were therefore not the right channels by which to achieve true cognition; indeed, doubt led him to conclude that all we can know for sure is that there is an *I* that doubts. Descartes helped to lay the foundation for the scientific approach that eventually displaced the speculative natural history and natural hermeneutics of the 17th century. But the phenomenological was also soon to be minimized in the study of nature.

The advent of biology changed the nature of the visual. No longer was the animate observed with the human eye; now measuring instruments were employed that quantified and measured all the elements of the animate that the sense could not discover. Biology then began to move in two directions: internally into the depths of the living organisms towards smaller and smaller parts, ending in DNA, biochemistry and atoms; and externally into the environment and complex ecosystems. The Danish philosopher Søren Kierkegaard (1813–1855) is quoted as saying in the first half of the 19th century that "where natural history is the science of the simplicity of everything, theology is the teaching of the complexity of everything"![88] But biology actually contains both a simplicity and a complexity; the study of the interior of the individual organism examines the inner simplicity of animates, and evolution biology examines their complexity.

The founding of positivism by the French scientist, August Comte (1798–1857), clarified this division in biology between researching inward and outward, between organism and environment. In his *Cours de Philosophie Positive* (1830–1842) "environment" becomes an independent subject area within biology.[89] The concept of environment is borrowed from Aristotle and Newtonian mechanics, which Comte employed in biology to denote the relation between animates and their surroundings. The division in biology is simultaneously a farewell to its last phenomenological element, the visual, for biology moves either in the direction of what is so small or what is so complex in time and space that either way it cannot be seen.

Among the biologists who followed the path into nature and the environment are the "new English naturalists" of the mid-19th century.[90] Their method consisted primarily of travel in order to study nature *in natura* instead of in collections and museums of natural history. In itself this was not a pioneering step; botanists such as Michel Adanson (1727–1806), Linné himself and their colleagues had also travelled in the 18th century. What was new was the nature of their observations. For them it was no longer a matter of collecting flora and fauna and classifying them in extensive systems, but of studying *biological communities*.

[88] Stangerup, Henrik: *Vejen til Lagoa Santa*, in *Naturens Historiefortællere*, Vol. 1, p. 79.
[89] Canguilhem, Georges: *Etudes d'histoire et de philosophie des sciences*, pp. 64–65.
[90] Ibid. p. 102.

Darwin, for example, was much preoccupied with the interaction of the various species, and his major work *The Origin of Species* (1858) abounds with reflections on the subject. Pursuing the great interest of his time, the development of nature, he linked his observations to the adaptation of the species to one another, to their habitats and to the progress of nature via natural selection. In 1869 Ernest Haeckel (1834–1919) named the area of biology that is concerned with the interaction between organisms and the environment "ecology". At this point ecology dealt with three areas: the relation of the *individual organism* to the biotic and abiotic environment, the changes in *populations* (e.g. the extinction or emergence of a species, its size and variation in numbers), and finally the composition, structure and metabolism (energy and matter) of the biological communities. The first level, the relation of the individual organism to environment often contains a phenomenological element in the form of observations that can be made directly through the senses in nature. In the two other cases the phenomenological part is less and the theoretical part correspondingly more. *Modelling*, which may be regarded as a synthesis of knowledge, is a central part of the theory of the ecosystem which is used to describe the attributes of ecosystems,[91] and which can therefore also be employed to make predictions. The description of allel development in a population using the Hardy–Weinberg Law on population genetics around 1908 is an example of modelling according to mathematical and statistical equations. The law can be used to demonstrate how natural selection can influence genetic composition, and as such it is both a quantifying of Darwin's idea of fitness yet at the same time a revival of Lamarck's idea of inherited abilities, if not at an individual level, then at a population level.

The area of biology that is externally trained towards the environment and the ecosystems, towards the great coherences in the nature that presents itself to our senses, therefore tends to be non-sensuous: the construction of theory builds on huge amounts of data, which can only be managed with the aid of computers and collected in models. Neither the flow of energy nor the food cycle is immediately available to our sense organs. The complexity of ecosystems extends far beyond what the senses can perceive: scientific knowledge of the lake that we can see, bathe in, drink from, sail on, smell, taste and feel belongs to another area than our senses. It has become an extraordinarily complex totality of genes, individuals, populations, societies, energy flows, food cycles, balances, adaptation and so on.

But ecology is only a subsection of biology; a steadily growing area of biological research is taking place in the laboratory. Where Darwin relied on direct observations in nature, the neo-Darwinists are attempting to explain evolution

[91] Jørgensen, Sven Erik: *Integration of Ecosystem Theories: A Pattern*, p. 47.

through genes. We cannot see DNA in nature, but only under the very best electron microscopes.

The microscope is above all others the instrument that has formed the theories of biology, for the science is built on the assumption that the complex can be explained from the simple, from certain fundamental, irreducible structures. Biology has moved from surface structures deeper and deeper into the animate, down to the smallest components, where it becomes organic chemistry and atomic physics. Biology has established a hierarchy over animate matter from molecules to organisms. The invention of the microscope by Anton van Leeuwenhoeks (1632–1723) and Marcello Malpighi (1628–1694) in the mid-17th century was the precondition for the investigation into the substance from which animates are made. Cognitively, biology became less and less phenomenological, and increasingly instrumental and theoretical. But it took time. Even though the microscope was invented long before Buffon and Lamarck, they did not use it. Observations of bacteria and sperm cells under the microscope did not develop into a generalized theory, and it never occurred to them to link the microscopic to the macroscopic, or to realize the fact, now obvious to us, that what we call the cell is the fundamental unit for bacteria (single-celled) as well as bigger, multicelled organisms, and that almost everything that is alive is composed of cells (though not, for instance, vira). At the end of the 1830s composite microscopes were developed with several lenses,[92] which improved dissolution. Different types of tissue could now be studied and everywhere the same structure was revealed; matter is made of small "cavities". The German biologist, Theodor Schwann (1810–1882), who in 1839 was the first to describe the cell structure of animals, characterized organisms as "states of cells", in which each cell can be compared to a citizen. With cell theory, biology finds its atom and the whole science is affected by the theory.[93] In the cell structure of matter, biology has found the unit that is common to all living beings, life's common denominator.

Around 350 years after Columbus's discovery of America in 1492, in a period when discoverers were on the way out, the microscope revealed a whole new world that was still unknown. The microscope increases the eye's dissolution of around 0.1 mm considerably. Two tiny pricks that are less than 0.1 mm from each other are seen not as two but as a single blurred prick. The light microscope, which magnifies the illuminated object with the aid of optic lenses, has a maximum dissolution of 0.2 μm (0.0002 mm), which corresponds to an improvement in eye power by a factor of 500. Through the light microscope most cells can be made visible, and larger structures such as chromosomes can

[92] Bonde, Niels: "Fra revolution mod evolution. 1800-tallets store anatomer and palæontologer". In *Naturens Historiefortællere*, Vol. 1, pp. 198–237.

[93] Jacob, François: *The Logic of Life*, p. 121.

also be observed in some cells. The electron microscope gives a maximum increase of 200,000 times in relation to the human eye. Its maximum dissolution is ~0.5 nm (0.0000005 mm). By comparison a hydrogen atom is ~0.1 nm in diameter and cannot be seen, though many larger molecules such as DNA can be observed with the aid of the electron microscope. It works by sending electrons through a specimen in which regions that allow the passage of electrons are light, while opaque regions rejecting its passage are dark. Greater magnifications in the light microscope and all magnifications in the electron microscope require very thin sections of what one wishes to observe (from 0.01 mm to 0.00002 mm), which makes it impossible to observe living cells in large magnification.

However, biology was not content to dwell on the cell and found yet another fundamental denominator: the gene. The concept of the "gene" was first used by the Danish natural scientist Wilhelm Johansen (1857–1927) in connection with the distinction he introduced in 1909 between phenotype and genotype.[94] As he imagined it, the gene was a "neutral" and "abstract" term for heredity; but with no concrete structure or formula at the time there was no basis in natural science for concluding anything about the formation of hereditary material. The geneticists, on the lookout for concrete and structured heredity material, had to wait even longer; until in 1944 Oswald Avery showed that heredity is to be found in the chromosomes (DNA). In 1950 Rosalind Franklin was able to show the first pictures of DNA taken with the aid of X-rays. In 1953 James Watson and Francis Crick described the structure of DNA as a double helix, and in 1965 the genetic code was deciphered by Marshall Nirenberg. But though heredity is well documented in its structure of molecular biology, the gene is still far from exhausted. The central dogma in genetics is that the hereditary information of an organism is to be found in its DNA in the form of genes, and that this information is transcribed to ribonucleic acid (RNA), which translates it into protein. At the core this formulation is still correct, but the picture is not complete without adding that research into this transcription, and translation gives an increasingly nuanced and complex picture of the process (called postgenomics). In other words, there is quite a long way from a lineal link between the DNA of the gene and its protein. This has led Kim Sterelny and Paul Griffiths to suggest that the definition of a gene should be as a DNA sequence along with its context.[95] The gene is thus theoretically a somewhat difficult entity, but with a major practical importance for scientific research. At the definition level, genes, like life, have no clear definition, and yet both are absolutely central to biological research. In the laboratory, as in nature, genes are not visible; in the technological process of making them so (which most of

[94] Johanssen, Wilhelm: *Elemente der exakten Erblichkeitslehre*, p. 124.
[95] Sterelny, Kim and Paul E. Griffiths: *Sex and Death. An Introduction to Philosophy of Biology*, p. 141.

all reminds one of the recipes in a cookery book), they are often subjected to different colour techniques such as fluorescence and radioactivity. The genetic material is made visible, but it is still not the genes that one sees, only what is colouring them.

Substance in biology is what separates animate from inanimate. The microscope revealed the differences between organic and inorganic on the organizational level. Molecular biology and biotechnology study the composition of the substance in living beings and define them first and foremost by matter, i.e. by their basic material. The new focus in our time on biotechnology has produced research results that emphasize convenience rather than difference between different living beings. Molecular definition of substance reveals that flora and fauna have so much in common that a division into an animal and a plant kingdom is somewhat artificial, at least in the view of many biotechnologists. Where biology could establish itself as an independent area of research 200 years ago by delimiting animate from inanimate, and thereby distancing itself from physics and chemistry, biology today is breaking down the differences between living beings, so that all the subject areas that are involved with animates are tending to gather under the single umbrella of life sciences.

Microbiology is posing some cognitively relevant questions at the moment because the same regularities in microphysics do not apply in our "mesocosmos", the term coined by the German biologist Hans Mohr to denote what man can detect with his senses (in vision that which is greater than 0.1 mm). The causality concept cannot be applied in microphysics. The theory of relativity speaks of another relation between time and space than the one we know from our everyday life. Our experience of space, time, substance and causality in the mesocosmos constitutes man's cognitive horizon – or following Mohr, our cognitive "niche" – which cannot immediately be used either in the solar systems of the macrocosmos or in the microscopic world.[96] With his sense and his corporality, man is born into the mesocosmos. In conclusion, the development of modern biology implies that nature is no longer accessible to direct sense perception, but must always be mediated via instruments.[97]

Let us turn to the models and theories of modern biology and how they emerge. This is particularly interesting considering the diminishing importance of phenomenological descriptions and the growth of the model approach. Biological models and theories are far from being purely phenomenological considerations of things as they appear to us; they are often affected or inspired by thoughts and ideas from other areas. Conversely, biological theory also serves as inspiration in other fields, such as sociology. An example of this that

[96] Mohr, Hans*: Natur und Moral. Ethik in der Biologie*, pp. 27–28.

[97] Cf. Seel, Martin: *Eine Ästhetik der Natur*, p. 22: "To the ill-equipped mind Nature is no longer conceivable; it is a theoretical construct of scholarship that has as its subject the foundations of human life, but not the living reality of human life".

has acquired classic status is to be found in the relationship between the texts of Malthus and Darwin.

François Jacob points out that Malthus's population theory was inspired by natural observations.[98] Malthus writes:

> Through the animal and vegetable kingdoms Nature has scattered the seeds of life abroad with the most profuse and liberal hand; but has been comparatively sparing in the room and the nourishment necessary to rear them.[99]

Malthus applies his observations of nature to develop his theory of the exponential growth of the population, which, with the inevitability of a natural law, creates famine because the food base is not increased at the same rate. He argues on the basis of a natural law in taking a theory about the order of nature and its development – even though it is admittedly his own – in order to apply it to society. Darwin, who was acquainted with Malthus's essay and acknowledges his debt to him, uses the theory in *The Origin of Species* to arrive at the principle of natural selection. In a letter from 1875 the philosopher and economist Friedrich Engels (1820–1895) comments thus:

> The whole Darwinist teaching of the struggle for existence is simply a transference from society to living nature of Hobbes's doctrine of *bellum omnium contra omnes* and of the bourgeois-economic doctrine of competition together with Malthus's theory of population.[100]

The origin of the theory is thus lost somewhere between observations of nature and social theory, and it oscillates between the two. But it does not end there, for Darwin's theory of evolution later inspires theories regarding the development of societies, giving rise to social Darwinism, which seeks to explain social development in the same way as the evolutionary development of the species.

The example illustrates the fact that biological theories often arise through inspiration from other sectors. The use of analogies and metaphors plays a significant role in the natural sciences. It also shows the two classic ways in which inspiration takes place: as anthropomorphism and by the so-called naturalistic fallacy. In the former case, human values and ideas are transferred to nature; in the latter, where the false step moves from *is* to *ought to be*, theories about nature are transferred to conditions in society.[101]

One cannot but ask what is the purpose of this mutual inspiration? why does it take place? why is it so easy to make these anthropomorphisms and naturalistic fallacies? There can hardly be any doubt that it happens during the attempt to create an intelligible order in the world; this is an interpretation that comes about by drawing analogies and narrating in order to understand. For order to

[98] Jacob, François: *The Logic of Life*, p. 186.
[99] Malthus, Thomas Robert: *An Essay on the Principle of Population*, p. 10 (Book 1).
[100] Schmidt, Alfred: *The Concept of Nature in Marx*, p. 47.
[101] Cf. Moore, George Edward: *Principia Ethica*.

be intelligible to us, there must be a link between things. The process of ordering sets things in relation to one another in both time and space.

In the light of the above, on the way biological theories and models are construed, it is tempting for a layman to regard biological theories and models as a kind of narrative representing possible explanations. In narrative research, narrative competence, i.e. the ability to understand narratives, is seen as a fundamental way for man to understand the world.[102] From a narrative perspective, when the biologist places various factors in relationship to one another and throws out a theory, the reader or listener attempts to understand it in the same way as one understands a story. The life sciences are converted into, or communicated as, texts and models of animates that are used in practice to control the living. The lay reader will often interpret the text on its technical qualities, for which as a rule he or she is not qualified, but the biological text will combine with a world of other information, accounts, news, views, expositions, prejudices and much more. The reader will therefore be liable to regard the biological theory as one narrative among many others. The French philosopher Paul Ricœur states in *Time and Narrative*[103] that narrative competence plays an important role not only in fiction but also in historiography. I believe the same to be true of biological theories: narrative competence prompts us to interpret biological theories as narratives, as stories. This is how we seek to establish an understanding of biology: as narratives that set up a coherent link in nature and place different factors in a relationship with one another.

If we accept this thesis for the moment, that most non-biologists relate to biological theories through narration and interpretation, what then are the consequences? The exploration of nature started in the sensuous and to that extent was a phenomenological and empirical activity to which all people had access through their senses. Through his sensory activity and corporality man was embedded in the animates' world. In Renaissance natural history nature was even embedded in social space.

But in two stages the life sciences are creating a steadily increasing distance between nature (which is also foodstuffs) on the one hand and sensuality, corporality and the social context on the other.

The first and crucial differentiation happens when the signs are no longer understood as the things themselves but as representations. This creates a distance between perception and the world, between man and nature.

The other important development occurred as the life sciences move away from human sensuality and into the organisms in order to find explanations in the ever-decreasing parts, in the microscopic: first in the organ itself, then in the cell, and finally in the molecule, and especially in the DNA. The area of biology

[102] See also Chapter Four for a more elaborated description of narrative research.
[103] Ricœur, Paul: *Time and Narrative*.

that has had the greatest impact in the 20th century has moved away from the phenomenological, from the macroscopic and sensuality, and has buried itself ever deeper in the tiniest elements of life. From noting that all life is constructed from the same basic unit called a cell (which means a "small room"), biology has moved on to the cell's interior with its macrostructures of chromosomes, cell nuclei, cytoplasma, mitochondria, cell membranes, endoplasmatic reticula, etc., which everyone can see through an electron microscope, but which are invisible to the naked eye. Under further magnification we can observe their surface, their structure and their components. We can note, for example, that cytoplasma is not just the dissolution of proteins but is interwoven with thin threads (3–6 nm in diameter). At the next stage of dissolution we can see the structures of macromolecules, e.g. in the larger proteins and DNA. In contrast, ecology and evolution theory are moving in the opposite direction, towards the increasingly larger, which is not open to sense perception either. The dominant timescale in biology is evolution, the perspective of which is so long that changes cannot be experienced by any single individual. Who has seen evolution – a biological species, a genotype or a gene?

Even though the French philosopher of science Gaston Bachelard (1884–1962) based many of his writings on modern science on the physics of the 1930s and 1940s, his reflections on the relationship between phenomenology and scientific theory also seem relevant to modern life sciences, when he writes:

> [A] truly scientific phenomenology is therefore essentially a phenomeno-technology. Its purpose is to amplify what is revealed beyond appearance . . . modern science has moved on to the project of constructing a world in the image of reason. Scientific work makes rational entities real, in the full sense of the word.[104]

Science no longer relies on phenomenological observations; rather it creates reality itself. He goes on to say:

> [B]etween sensory knowledge and scientific knowledge there is a gap. Temperature is seen on a thermometer, one does not feel it. Without theory, one would never know whether what is seen and what one feels correspond to the same phenomenon.[105]

According to Bachelard, modern science is not concerned with describing realities, but with the realization of the rational.

The instruments of scientists cannot be considered mere extensions of their senses and perceptions; rather they are considered new organs with the specific purpose of reducing or completely eliminating the use of the senses within science; in so doing, they become pure and objective receptors. Modern science has been emancipated from perception; science works not with perceptions, but

[104] Bachelard, Gaston: *The New Scientific Spirit*, p. 13.
[105] Bachelard, Gaston: *The Philosophy of No*, p. 9.

with instruments. Bachelard emphasizes the importance of the relationship between instruments and theory by saying that "instruments are nothing but theories materialized. The phenomena they produce bear the stamp of theory throughout".[106] Perception has been reduced to the reading of instruments, and with the measured index figures new theories are constructed.

When the almost de-phenomenologized and de-sociologized life sciences are setting the agenda for our relationship to food, there cannot but be problems, for their findings harmonize neither with our sensual and corporal experience nor with our existential life situation. The narratives of animates that the life sciences present are positivist and materialistic, and there is nothing wrong in that, although it should not become too dominant as this allows precious little room for any form of food ethics.

[106] Bachelard, Gaston: *The New Scientific Spirit*, p. 13.

CHAPTER THREE

THE STORYLESSNESS OF FOOD

We saw in Chapter Two how the emergence and development of the life sciences brought about changes in our understanding of nature and thus also of food. Even though these sciences have made a significant contribution to increased living standards and a safer appraisal of food in the Western world, they can nevertheless be criticized for the intellectualization and distancing they have added to our relation to food. This criticism is important, for it can show us some of the difficulties involved in relating ethically to food.

Nor is it only the life sciences that have influenced our relation to food. Their growth has also been accompanied by a more practical development in agriculture and food manufacture, which is the subject of the first section of this chapter. These too have contributed to the distancing of the consumer from the production history of food.

The second section of the chapter deals with the consumers' attitude to food and takes the form of a critique of their lack of opportunity for relating ethically to the circumstances that dominate modern food production.

1. THE HISTORY OF INDUSTRIALIZATION OF AGRICULTURE

The general increases in agricultural yield from the 16th century through to today are often linked to the industrial revolution. The central question concerning the *agrarian revolution*, which in many ways has involved the industrialization of agriculture, is how agricultural production has been able to multiply to such an extent that it can feed the rapidly growing population, yet simultaneously release labour power to industry. For this to happen, the total yield had to grow and the yield per labourer even more so. Whereas the total yield can grow through an increase in labour effort without the use of new methods, the yield per labourer focuses on increased efficiency and the rationalization of agriculture. This is why it is the yield per labourer that best describes the industrialization of agriculture.

There are many factors that have contributed to such a seismic increase in agriculture. If by industrialization we understand an increased efficiency of labour power – with the division of work processes into separate and specialized units, as in the textile industry – we can say that many of the initiatives that have helped to increase agricultural production since the 16th century were not typical forms of industrialization. The increase in agriculture production, for instance, came about through the cultivation of new crops. Clover was one such, and was from the 16th century onwards increasingly used in European agriculture; indeed by virtue of its ability to absorb nitrogen from the air and thus increase soil fertility, it helped to boost the yield significantly. But clover did not reduce the need for manpower; nor did it help to specialize farming; perhaps even on the contrary. The use of clover cannot therefore be regarded as typical industrialization; more precisely it should be described as the employment of a biological resource with the aim of increasing the agricultural yield.

The industrialization of agriculture differs from that of trade and craft by not only taking the form of a special organization of labour power but also by being *a particular organization of nature*. In order to understand the industrialization of agriculture we should therefore examine how nature's own processes are utilized so that we may increase the yield. But there are of course more typical forms of industrialization to be found within agriculture, such as the substitution of tractors for horses in the 1950s. This move bears much more the stamp of industrialization, partly because it increases the yield per man-hour and partly because horses are replaced by mechanics and fossil fuels.

Let us return to the growing population in Europe. From around the 16th century the number of people in Europe began to rise again after a period during which the plague had decimated the Continent. Between 1650 and 1800 the population of Denmark, for example, rose from ~550,000 to 925,000[107] and in the following century the increase was even greater, since the epidemics of the past, plague, malaria, tuberculosis and small pox, were diminishing. According to the Danish historian, Thorkild Kjærgaard, the disappearance of these diseases (with the exception of small pox) was due not only to the progress of medical science but even more so to changes in cultivation and farming practices and the consequent transformation of the landscape.

The plague is a good example of the link between the transformation of the landscape and the disappearance of a disease. The received wisdom is that the plague disappeared from Europe as a result of improved sanitation and hygiene, which led to a curtailment in the number of rats. This is indeed part of the story, but it is far from the whole truth. According to Kjærgaard, it was rather a result of woodcutting. The felling of timber reduced the number of rodents, which are the actual carriers of the plague bacillus via fleas to humans. The lack of wood also meant that stone began to be used instead for housing; stone houses are less attractive to rats and therefore reduce their contact with humans. Kjærgaard

[107] Kjærgaard, Thorkild: *Den danske revolution 1500–1800*, p. 18.

believes that changes in farming practices were of far greater consequence than is generally realized, and that the significance of new methods of cultivation for epidemics via changes in the landscape was an unforeseen consequence that enabled the population to increase.

Kjærgaard's point is that the interplay between agricultural practice and population numbers may be more complicated than is normally accepted. Agricultural practice is not only important for whether there is enough food or not but also, as we have seen, for whether epidemic diseases continue to ravage the population. This suggests that the relation between population growth and agricultural production is not merely causal; it is a simplification to say, as is often done, that a growing population required more food or that increased yields from agriculture facilitated a rise in population. Of course a growing population needs more food, but it was changes in agricultural practices that led to changes in the landscape which made a growing population possible.

In industrial countries the yield per acreage, known as the acreage production, has risen substantially since the 16th century.[108] Malthus's famous claim of 1798 in *An Essay on the Principle of Population*, that an exponential population growth could not always be counteracted by increases in agricultural yield, has proved false. It is also clear that the yield per labourer has increased proportionately, particularly in the 20th century. The number of people employed in primary agricultural production today is very small compared with that in the past.

Historians have made many attempts to locate the agrarian revolution to a defined period and to specific initiatives such as the introduction of particular crops, a change in crop rotation or the innovation of machinery.[109] Some even propose three separate agrarian revolutions at different times, each characterized by different initiatives.[110] But it is probably more reasonable to speak of a series of gradual changes that slowly led to an increase in yields and profits and in line with greater efficiency.

In his work on the diffusion of new technology in agriculture, Hans Jørgen Winther Jensen has divided the changes that led to increased yields into four categories.[111] I append the most important elements in each group:

(1) **Objects of labour**: soil, livestock, crops and water.
(2) **Means of labour**: tools, machines, pesticides, input of new resources such as energy, fertilizer, etc.
(3) **Methods of labour**: soil treatment, crop rotation, breeding, draining.

[108] Cf. Overton, Mark: *Agricultural Revolution in England*.
[109] Cf. Beckett, J.V.: *The Agricultural Revolution*.
[110] Holderness, B.A.: "Apropos the Third Agricultural Revolution: How Productive was British Agriculture in the Long Boom, 1954–1973?", pp. 68–87.
[111] Jensen, Hans Jørgen Winther: *Diffusionen af ny teknologi på bondebrug i Danmark 1800–1915*, p. 2.

(4) **Organization of labour**: marketing and sale, work form (e.g. cooperation between several farmers, length of working day, division of work, etc.), forms of ownership, acquirement of expertise, professional skills, etc.

Objects of labour (which in this case are all natural elements) are the fundamental elements that agriculture works with: what in a general sense belongs to farming. Traditionally these were sustained by agriculture itself, and were for the most part independent of outside input. The farmer bred his own livestock and his own crops and only occasionally bought new animals. As we have seen, new plants were introduced, but as soon as these had become part of a farm, they too were reproduced by the farm itself. Soil and water were more or less a gift of nature, though soil productivity could be improved by the resources that lay within the farm.

Primarily the farm was self-sufficient, since most people lived in the country and worked in agriculture or housekeeping. The distance from field to fork was a short step; the link was simple and tangible. As the farmer and the village saw it, agriculture was more or less a closed circle; there was only minimal input from the surrounding society, produce was consumed *in situ* and the waste returned to the earth.[112] No other resources were employed than the farmer's work – very rarely did new crops such as clover appear – and only minor amounts were sold elsewhere.

The industrialization and modernization of agricultural tools began with the onset of this cycle. In place of the almost closed circle came a linear production sequence that can be described as:

$$\text{input} \Rightarrow \text{agriculture (production)} \Rightarrow \text{consumption} \Rightarrow \text{waste}$$

This linear production chain is characterized by the waste not being recycled as input for new production, but being allowed to accumulate. During and after the industrialization of agriculture, the natural entities were no longer local but increasingly premanufactured and processed products. The soil is an example of this.

The preservation of the soil's fertility has always been essential for farmers, the guarantee that they could harvest year in, year out. Various methods have been tested to retain fertility and avoid exhaustion, and common to all of these has been the fact that they have rested on the plants' own ability to recreate that fertility. It was appreciated that the soil regularly needed to lie fallow in order to regenerate. In his book *Grundsätze der rationellen Landwirtschaft* (1809–1812), Daniel Thaer (1752–1828), Professor at the University of Agriculture in Berlin, supported the theory that had long existed that plants

[112] The circle was not quite closed; at times the forests and land were overexploited to the point of exhaustion and erosion.

nourished themselves through water and humus. Humus was regarded as crucial and the plants themselves were believed to amass the humus, e.g. when lying fallow; animal manure was also known to improve soil fertility. However, in the course of the 19th century ideas and theories of soil processes and plant functions gradually changed. The humus theory suffered a decisive blow with the publication in 1840 of Justus von Liebig's *Die organische Chemie in ihrer Anwendung auf Agricultur und Physiologie*. Liebig confirmed (as others had before him, though without effect) that plants absorbed minerals from the earth, not humus, and could grow in the soil without it. Liebig proposed that inorganic fertilizer could be used instead of animal manure.[113] He therefore replaced manure with nitrogen and other minerals, signalling the arrival of artificial fertilizers in American and European agriculture. Liebig noted that waste products from a number of industries contained minerals that could be used as fertilizer, and as early as 1843 the production of super-phosphate began in England.

For over 10,000 years livestock and crops have been the farmer's own production sphere. Before the arrival of scientifically aided agricultural produce, the farmer chose the best individual specimens to produce new offspring. One of our earliest sources is Aristotle, who describes how animal breeders, by examining the animals' teeth, knew the difference between old and young "quadrupeds" – which formed a separate group in his classification, as opposed to six-legged insects, for example.[114] In the second half of the 19th century the first steps were taken to systematize and institutionalize agricultural produce. Breeding became a separate business, and preceded agriculture. This tendency took effect especially in the plant world, whereas it varied for livestock breeders. Cows are generally bred from the farm's own stock, though with the help of scientifically chosen stud bulls. Modern chicken production on the other hand has followed in the footsteps of plant breeding. Five major firms produce more or less all the broiler chickens in Europe today.

With industrialization the soil, animals, plants, manure and water all change status. Formerly these were more or less fixed and defined by local conditions. They were given natural elements. With industrialization they become an increasingly standardized *input* of agriculture – its *biocapital* – in the sense of providing capital interest. The unique feature of this capital is that it is biological.

The development and application of **means of labour and tools** to replace labour power was dependent on the progress of industry, for only industry was able to create the new and complicated tools that the local smithy could not forge. In the 18th century wood continued to be the most common raw material for agricultural implements such as ploughs, harrows and seed sowers, but

[113] Brock, William H.: *Justus von Liebig. The Chemical Gatekeeper*, p. 148.
[114] Aristotle: *History of Animals*, 578b5–578b10.

this was replaced by iron in the 19th century.[115] The introduction of new machines to agriculture was remarkably slow compared with industry, the real labour-saving machines only appearing in the 20th century.[116] The reason for this tardiness is to be found not least in the widespread availability of labour in the countryside, making it so cheap that it was not profitable to invest in new machines.

With the changes in the objects and means of labour came changes in **methods of labour**. As mentioned, processing moved out of the farm and into the factory in the course of the 20th century. Crop rotation and animal husbandry, two of farming's most important methods, also underwent great changes. In industrialized agriculture, crop rotation since the 1950s has been rendered well-nigh superfluous through the use of fertilizer and pesticide combined with highly processed strains. Instead, the farmer can grow the same crop in the same soil year after year. And the arrival of the automatic feeding plant and barn cleaner have made it possible to keep large numbers of livestock in stable all year round, so that the animals do not need to graze.

In 1817 the economist David Ricardo (1772–1823) introduced the concept of *comparative advantage*. By way of example, Ricardo claimed that both England and Portugal would achieve the greatest benefits if they produced what they were best at. England ought to produce textiles and Portugal wine, and then trade these articles with one another. On the other hand, if both countries were to produce both articles, the advantage would be less. The idea of the comparative advantage still plays an important part in agricultural development, and takes the form of a comprehensive specialization: in 1968, 75% of all Danish farms kept both pigs and cattle. By 1999 this figure had fallen to a mere 10%.[117]

The **organization of labour** in agriculture was linked to developments within a number of other areas such as the economy (mercantilism in the 18th century), the separate independent areas of natural sciences like chemistry, biology and nutrition science, and the contribution of engineering to the development of machines and the infrastructure. But let us first consider the economic advances.

"Physiocrats" is the term for a number of economists from the 18th century with their roots in agriculture who elaborated a theory of the emergence of material values and material wealth. They were for the most part themselves landowners, though not "yeomen", an ownership form of a smallholding that emerges in 18th century Europe to replace feudal society. The physiocrat Quesnay writes:

[115] Cf. Moberg, Harald: *Jordbruksmekaniseringen i Sverige under tre sekel.*
[116] Beckett, J.V.: *The Agricultural Revolution*, pp. 25–28.
[117] *Tal om landbrug*. Landbrugsrådet, København, 2000.

> The fruit I am hungry for, which I pick and eat, is a *commodity* presented to me by nature; there can be no *wealth* unless the fruits on my tree are sufficiently numerous to exceed my appetite. Even then, someone else must be hungry and require those fruit of me.[118]

The use of *commodity* here is interesting. According to the physiocrats the price of these *commodities* is fixed in the actual exchange of the articles, in the interplay between the supply and the demand. They regarded consumption in particular as determining the value of the article, and agriculture differed from all other forms of production by being the only sector in which an increase in value due to production was not attributed solely to the producer's ability, for when the land was cultivated, it provided a wealth of essentials greater than the farmer himself required; in other words, he had a secret partner, the Creator of nature, who was Himself the actual producer of all wealth and all good gifts around us.[119] Among 18th century economists, including Adam Smith (1723–1790), the general view was that any accruing interest was an expression of the specific fertility of the soil; which is why this period sees so many analyses of the origins of wealth.

What is of note is that the physiocrats argued that the fixing of a value only takes place in regard to the surplus that remains after the farmer has covered his own needs. They attributed this profit to the "rent of land", the land being regarded as a capital asset providing interest and as the ultimate source of all material values.

On the other hand, the farmer was no different from the industrial worker in being seen as a labour tool.[120] In the 18th century it was food, clothing and housing – the working man's needs – that were the yardstick for the market price, with soil and corn being considered particularly dominant.[121]

The influence of the physiocrats on the agriculture of their day was apparent in two areas. First, if wealth came of its own accord from the richness of the land, there was little incentive to improve the land and make agriculture more efficient. Any surplus was the gift of the Creator. Second, the physiocrats pointed out that it was possible for agriculture to produce itself – with divine aid simultaneously creating a surplus – and this brought agriculture into the market economy. The physiocrats mixed profane economic considerations with sacred motives in this "religious mercantilism", arguing that agriculture per definition created a surplus, for that was how God had furnished nature. It followed that this surplus should be the object of trade: food should be a commodity.

118 Quesnay, quoted from the article "Homme" in Daire, Eugène: *Les Physiokrates* (1846), p. 42. Here quoted from Foucault, Michel: *The Order of Things*, p. 192.

119 Mirabeau, Marquis de: *Philosophie rurale*. (1763), p. 37. See Foucault, Michel: *The Order of Things*, p. 194.

120 Mirabeau, Marquis de: *Philosophie rurale*. (1763), p. 8. See Foucault, Michel: *The Order of Things*, p. 192.

121 Foucault, Michel: *The Order of Things*, p. 222.

The physiocrats' theory of the bounty of nature did not go uncontested for long. At the beginning of the 19th century, agricultural abundance soon gave way to a shortage in the face of the burgeoning population, whose numbers, as Ricardo's contemporary, Thomas Malthus, pointed out, made it increasingly difficult to supply food. Foucault writes:

> *Homo oeconomicus* is not the man who represents his own needs to himself, and the objects capable of satisfying them; he is the man who spends, wears out, and wastes his life in evading the imminence of death.[122]

The struggle for survival became a main theme in biology too. Darwin wrote a whole chapter on the "Struggle for Existence" in *Origin of Species*.[123] A radical rethinking of agriculture took place: its purpose now was to increase yield. And to that end the natural sciences were taken into use.

But foodstuffs were not only to be produced but also required transportation to the consumers. The arrival of the railways in the middle of the century improved the infrastructure significantly, and it became possible for many more farming areas to export and import foodstuffs. Such changes were tangible, for with self-sufficiency the overriding interest is versatility, whereas with the market economy the prime interest is in specialization, since this offers greater opportunity for efficiency, sales and export. With self-sufficiency there is no reward for producing more than is needed, whereas in a market economy the more one produces, the more one earns.

The transition to a market economy also involved a change in work forms. More food was required for the rapidly increasing city populations. Yields should be increased. As we have seen, mechanization did not really make its mark until the 20th century, so in the early days yields could only be increased by intensifying the labour effort. Thorkild Kjærgaard estimates that the working hours in society rose between 1500 and 1800 by 50%, and even more so in agriculture.[124] The number of working days, and their length, was increased, and in the course of the period the rural population had to accustom itself to long hours of regular work.

Specialist literature on farming has been around for hundreds of years in Western Europe; indeed there are even ancient Greek writings on the practices of good farming.[125] Together they have provided an important communication

[122] Foucault, Michel: *The Order of Things*, p. 257. As demonstrated above, this is a view that influences economic theories, but is also to be found in biology; cf. Bichat: "life is the sum of the functions that oppose death".

[123] Despite the general opinion, at the time of publication of *Origin of Species*, Darwin was unaware of the development theories of Herbert Spencer (1820–1903). The book was published in six edited versions, in the first two of which Darwin employs the concept of "natural selection", and it is only in the later editions that he uses Spencer's concept of the "survival of the fittest".

[124] Kjærgaard, Thorkild: *Den danske revolution 1500–1800*, p. 142.

[125] See Isager, Sigen and Jens Erik Skydsgaard: *Ancient Greek Agriculture*.

system on farming and its regular renewal. But it is not until the mid-19th century that an actual science of agriculture was established, with the foundation of the first agricultural colleges (also known as farm institutes). Agricultural science built on the natural sciences of chemistry, physics, mathematics, statistics, geology, anatomy, zoology and botany.

Summarizing, we can say that industrialization brought a general *homogenization* of agriculture, which simultaneously marked a *severance* from the local framework. The Dutch professor of rural development, Jan Douwe van der Ploeg (b.1950) uses the phrase *dislocation and disconnectedness* about this development[126] to describe the ending of the embeddedness of rural social relations and their anchorage in time and space. With industrialization, agriculture becomes independent of local conditions. Van der Ploeg enumerates four areas where homogenization and dislocation took place.

First, there is the dislocation from the land, nature and ecology. The genetic variation of plants and animals decreases because only a few strains and races are used, while the form of production is standardized according to scientifically based optimization of output, and the processing of foodstuffs is standardized to meet industry demands for mass production. At its most extreme, "dislocation" is the total control of climate and soil, as typified by the greenhouse.

Second, fewer and fewer tasks were the preserve of local manpower. Instead they were distributed to external areas. Agriculture receives an increasing amount of processed inputs. Automation involves the standardization of work, which is then no longer linked to local factors, but is adapted to external conditions.

Third, agriculture becomes dislocated from the end product. Agriculture supplies goods to the food industry, which processes them and passes them on to the wholesaler, who in turn delivers them to the shop chains. The direct link from farmer to consumer is broken.

Lastly, agriculture loses its connection with a family structure. Families may continue to inhabit farms, but this is no longer of significance for the work of agriculture. Through industrialization, the traditional agrarian form of life is lost, to be replaced by the theory of productivity. From now on agriculture is regarded as a production unit on a par with other industrial businesses.

2. FOOD SCIENCE AND GASTRONOMY

The intellectualization of food reveals itself in the arrival on the scene of two subject areas in the second half of the 18th century: food science and gastronomy. Both can be seen as offshoots of the contemporary interest in understanding the nature of man. Towards the end of the 18th century it is mankind itself

[126] Van der Ploeg, Jan Douwe: "The Reconstitution of Locality: Technology and Labour in Modern Agriculture", pp. 19–43.

that becomes the object of the sciences. As we have seen, the consumption of food occupies the border ground between the self and the other, and it is therefore of particular interest to understand the relationship between man and the rest of nature. The study of digestion had already begun in 1752 with René Ferchault de Réaumur's *Second mémoire sur la digestion*, in which digestion is described as the breaking down of food followed by fermentation. In the second half of the 18th century, Antoine Lavoisier (1743–1794) also studied the functions of the body from observations of the combustion of food, the passage of energy through the body and the separation of waste products.[127] As we have seen, biology was directed more and more towards the specifics of life: what distinguished the living from the non-living. The metabolism was an area that clearly separated the one from the other. Organic chemistry and biochemistry studied the composition of the substances that pass through the living, while their path and transformation in the organism was examined. Many biochemists such as Liebig understood the vital force as the capacity to break down food and transform it into other substances, thereby constructing new bodies.

In the 18th century food science was still physical; it was the pumping of blood and the metabolism as it appears in the digestion process that were studied.[128] But from the 18th century the focus turned perceptibly towards the individual substances. Here began the chemical analysis of food, its division into organic substances: proteins, vitamins, sebacic acids, fibres, carbohydrates, lipids, etc. The fascination with the new science led Friedrich Accum to describe the culinary art in 1821 as chemistry: "The kitchen is a chemical laboratory ... the boilers, stew-pans, and cradle spit of the cook correspond to the digestors, the evaporating basins, and the crucibles of the chemist".[129]

Food science was not just a basic science but was also an extremely practical one. As applied science, the preservation of foodstuffs was one of its central articles, important partly because the climate of northern Europe made it impossible to produce foodstuffs all year round and partly because good methods of preservation also meant that too much wastage from poor storage was avoided. In northern Europe, April was known as the dead month, being the period where stocks were at a minimum and spring vegetables not yet available. Better preservation of foodstuffs was therefore a way to avoid starvation in April. Food hygiene was another area of the new science. Hygiene was important for a successful preservation of foodstuffs as well as for combating diseases. "There is death in the pot", wrote Accum in *A Treatise on*

[127] Jacob, François: *The Logic of Life*, pp. 42 and 83.
[128] Mellemgaard, Signe: *Kroppens natur*, p. 263.
[129] Accum, Friedrich: *Culinary Chemistry*, Preface. This is not so very far from the view of a modern nutrition expert. For example, Ernest Vieira writes: "Foods are made entirely of substances that, in the pure form, can be described as chemicals or chemical compounds" *Elementary Food Science*, p. 185.

Adulterations of Food (1820), which is a discourse on the various forms of foodstuff poisoning.[130]

Ernest Vieira also emphasizes the practical content of food science: in 1795 Napoleon Bonaparte asked Nicholas Appert (1749–1841) to help to make his troops less dependent on local supplies in wartime, since it made his soldiers less effective. Appert's solution to the problem was tinned food! Drawing on Spallanzani's results among others, Appert developed the first practical method of thermal preservation and was the first to use tins as packaging.[131] There is no doubt that from the very start of food science the conservation and durability of food played a crucial role.

Another way of putting food science into practice was invented by the German chemist, Justus von Liebig, who was an important figure in the development of the food industry. His theories that the essential nutrients in meat were to be found in its liquids led to the development of "Extractum Carnis Liebig". The product, mass-produced from South American beef by Liebig Co., was sold throughout the world and praised in its day for its nutrient value, but today it is regarded as ineffective. Liebig's studies of flour bolting in 1868 demonstrated the large loss of nutrients during the bolting, and he experimented with different forms of baking powder, though without achieving any commercial success. On the other hand, his studies of the composition of breast milk were employed in the mass production of breast milk substitutes and baby food. It was on the basis of Liebig's theory of nutrition that Henri Nestlé started a production of powdered milk in the 1860s and Julius Maggi and Carl Knorr began to produce concentrated soup in the 1870s and 1880s. Liebig Co. still exists as part of the food giant Unilever. Drawing on his scientific studies and assuming the availability of industrially produced foodstuffs, Liebig published a number of recipe books in the 1870s, including *Improved and Economic Cookery*, in which he stressed the nutrient value of food. Nor was he blind to the possibility that the recipes might improve his sales, since practically all the 100 recipes listed Liebig's Meat Extract.[132]

Today food science denotes the field of study concerned with food under production, processing, digestion and excretion. This is not the place to present a comprehensive account of the many areas of food science, but Table 3 gives an impression of its complexity and many branches.

The different areas cannot be separated in practice. For centuries, indeed millennia, the addition of salt, sugar, vinegar, spices, honey and berries, as well

[130] Accum, Friedrich: *A Treatise on Adulterations of Food*. 1820. Reprinted in Morton, Timothy: *Radical Food. The Culture and Politics of Eating and Drinking 1790-1820*. Routledge, London, 2000, Vol. II, pp. 287–383.

[131] Vieira, Ernest: *Elementary Food Science*, p. 3.

[132] Brock, William H.: *Justus von Liebig. The Chemical Gatekeeper*, pp. 215–249.

Table 3. Overview of the most important areas of food science

Area	*Keywords*
Nutrients	Carbohydrates (sugars and polysaccharides), proteins, vitamins (soluble in water or oil), lipids (fats, oils, lipids, wax, etc.), fibres, minerals, secondary substances, etc.
Enzymes	Proteases, oxidases, lipases, etc. often form part of the breaking-down process, but can also be used in processing.
Composition	Anti-caking agents, emulsifiers, stabilizers, thickeners, etc.
Additives	Substances that are not in themselves foodstuffs, but are added for a particular purpose.
Taste	Enhancement, preservation, concealment, addition. For example, lipases can add a particular flavour to cheese and powdered milk.
Conservation	Heat treatment, curing, refrigerating, packaging, additives, etc.
Toxicology and hygiene	• Food infection: pathogen organisms such as Salmonella bacteria. • Food contamination: the infected organism produces a toxin. • Food authority requirements.
Processing	Fermentation, storing, enzyme technology, etc.

as the process of curing, has been among the treatments employed to preserve food and to give it a particular flavour.

Even though food science has taught us to distinguish between many different substances in food and has provided better knowledge of their effect on the body, it has also presented us with a host of unanswered questions. Breast milk, for example, is thought to contain several hundred chemical substances, but only a few of their chemical structures are known.[133] There are tens of thousands of secondary substances in food that will require an enormous effort to chart; and an even greater effort will be needed to understand their significance.

Yet food science has contributed to radical changes in foodstuffs and our relation to them. Food science has introduced methods of standardizing food but also in some areas led to increased diversity. Many foods have become standardized because differences caused by time and locality have been smoothed out. Cheese is a product whose taste and consistency used to depend very much on time and place, for the animals and their milk were affected by place (soil, plants, climate, etc.), season (e.g. a particular cheese would be made from the milk of the cows grazing in early spring) and the local processing. Standardization is of course linked to the fact that the mass production of foodstuffs cannot include such a multiplicity of factors, but it also has to do with the

[133] Vieira, Ernest: *Elementary Food Science*, p.185.

development of food science as an international discipline with standard procedures and standard demands. However, food science has also contributed to the development and dissemination of a diversity of food products, e.g. all the new products based on milk.

Food science has elbowed its way in between the field and the kitchen and few things have been left untouched in its path. For instance, plant-extract proteases can be added to meat to make it leaner, either by giving the animal an injection before slaughtering or in powdered form distributed over the butchered meat.[134] Enzymes are often added to flour to make it more digestible for humans. Increasingly few foodstuffs are what consumers believe them to be, and particularly in the case of genetically modified food (GMF), the scientific element is patent. In 1990 the US Food and Drug Administration (FDA) in collaboration with WHO recommended the worldwide use of comparative data when assessing novel foods.[135] Novel foods were to be compared with traditional foods. In 1993 the Organisation for Economic Cooperation and Development (OECD) designated this principle for the analysis of GMF *substantial equivalence.*[136] Much has been done since to make it operational, but the principle has also been severely criticized, for though there may be similarities at significant points, this does not necessarily mean that food safety is secured. One important problem is the difficulty of comparing foodstuffs chemically at all, since the differences in content of the chemical substances vary according to the farming methods and conditions. Even potatoes from the same field can show considerable chemical variation.

Food science has enabled consumers to choose articles of food on the basis of their chemical composition. This has had particular appeal for those who consciously make nutrition part of their general health. Food and diet have dominated the food debate since the 19th century. Food places man under a specific morality, namely a discipline of the body and a self-control: "I must eat that" – and especially, "I *mustn't* eat that". In contrast to seasonal limits on available food or the Old Testament regulations as to what could or could not be eaten, the regime of food science works through the formation of norms: eat your way to health. However, this is neither a law nor a directive to be obeyed, the violation of which will be punished. The norm may come into being through external forces, as with seasonal limits, but it differs by establishing itself *internally*, as an "inner" norm.[137] In contrast to the norms of seasonal limits, it requires self-control to follow the norm of eating healthily. The punishment for

[134] Ibid. p. 128.

[135] WHO: *Strategies for Assessing the Safety of Foods Produced by Biotechnology*. Report of a Joint FAO/WHO Consultation, WHO, Geneva, 1991.

[136] Nordic Council of Ministers: *Safety Assessment of Novel Food Plants. Chemical Analytical Approaches to the Establishment of Substantial Equivalence*, p. 11.

[137] Lupton, Deborah: *Food, the Body and the Self*, p. 74.

disobeying is self-inflicted: guilt, illness and a strain on the public health service. In this way food science offers a scientific alternative to gastronomy, to taste and hedonism, in the choice of food.

Gastronomy too has had a determining influence on the relationship to food. As a concept, it can be traced back to 1623,[138] but around 1800 it takes a decisive turn. For as befitted the times, there was an initiative to turn gastronomy, which was otherwise concerned with pleasure, into a science: the science of taste.

"Gastro" comes from the Greek for "stomach"; gastronomy therefore has to do with the rules and regulations of the stomach. As with the other natural sciences, the science of gastronomy seeks to discover and describe regularities, and this is what the French lawyer, Jean-Anthelme Brillat-Savarin, set as his goal. Inspired by contemporary physiological studies in biology he wrote *Physiologie du goût* (*The Physiology of Taste*) in 1825, developing a gastronomic method that he called *éprouvettes gastronomiques*, or gastronomic tests, the purpose of which was to separate the senseless proletariat from the real gourmets, the true hedonists, who could appreciate a well-cooked meal. Brillat-Savarin defined the gastronomic tests as follows:

> By gastronomic tests we designate dishes of recognized savour and of such acknowledged excellence that nothing more than the sight of them will awaken, in well-balanced man, all his gustatory powers; as a result, anyone who in the same situation shows neither the flash of desire nor the glow of ecstasy can rightly be set down as unworthy of the honours of the gathering and all its accompanying delights.[139]

In a way it is sensory science that comes into existence here as a professional discipline. This is underlined by the fact that Brillat-Savarin in his book proposed to establish an *Académie des Gastronomes*, though this was not realized until 1928. As we have seen, Brillat-Savarin held up the natural sciences as his model, and thus the visual element came to play a special role in gastronomic science, a status that was then transferred to the area of taste. Taste was to be transformed into something visual, to be read on the face of the diner expressing his appreciation or other impressions of the food as he ate.[140] The face cannot hide the emotion of the inside, and was therefore seen as being more truthful and objective than words. Brillat-Savarin wished to make gastronomy objective through the use of physiognomy and phrenology. The same thing happens today in a different form, though it is not the diner's face one reads; nowadays the attempt is made to connect particular taste impressions and specific pictures of the foodstuffs, such as the appearance of so-called MRI pictures of foods. With the aid of spectroscopy, whereby the food is irradiated by a number

[138] Onfray, Michel: *La raison gourmande. Philosophie du goût*, p. 106.
[139] Brillat-Savarin, Jean Anthelme: *The Physiology of Taste*, pp. 173–174.
[140] Onfray, Michel: *La raison gourmande. Philosophie du goût*, p. 122.

of specific wavelengths outside the capacity of the human eye, we can via reflection and transmission determine the physiognomy of the foodstuffs and their chemical composition.[141]

Rabelais and Montaigne called gastronomy *science de geuele*,[142] the science of the mouth, an expression that in a subtle way also points to the importance of the articulation of taste. The link between pleasure and knowledge, i.e. between taste and verbal expression, was at the heart of gastronomy. This articulation was particularly the work of Alexandre-Balthasar-Laurent Grimod de la Reynière (1758–1838), a rich man's son and Parisian eccentric. As a result of his deformed hands and eccentric character, he was banished from Paris at a young age by his parents, but on the death of his father in 1792 he returned to the family's mansion on the Champs Elysées. Here he became renowned for his exclusive, bizarre and extravagant dinner parties, which amounted more or less to happenings. For instance, some dinners took the form of pseudo-funerals for the participating guests. Another happening was his "Déjuners philosophiques ou semi-nutritifs". Grimod was publicly forbidden from reviewing plays of the time, probably because his reviews were too excessive, but this gave him the idea of reviewing food instead. He therefore gathered around him the first panel of arbiters of taste in the world, before whom the Parisian chefs could present their dishes for appraisal. The food was reviewed in the periodical *Almanach des gourmands*, which Grimod started in 1803. However, the periodical also included issues on ordinary food, reviews of food stores in Paris and so forth. This was written in a style that today reminds us of our contemporary journalism. It became the birthplace of all gastronomic writing, with its comments on the gastronomic scene and its extraordinary mixture of technical advice and criticism coupled with poetry and polemic.[143] It was an articulation that even reserved the right to name the dishes in relation to their method of preparation, their techniques and their place of origin: *Lapins à la vénitienne, chapons panés à l'anglaise, homards à la purée d'oignons parfumée aux clous de girofle, figues et raisins aux deux épices* . . . . The periodical was a colossal success, but it was not long before it met with protests and accusations of being prejudiced and partisan, and under the threat of court action Grimod was forced to close down the venture in 1812.

Gastronomy was also closely linked to table manners. Since the 16th century Western civilization has experienced considerable change in this area, described by Norbert Elias as *civilization* (cultivation) in French and *Bildung* (culture) in German. Cultivation and culture came to represent the opposite of uncivilized and natural. Manners and culture tamed or suppressed natural man.

[141] Munck, Lars: "Levnedsmiddelteknologerne overtager astronomernes værktøj". pp. 43–49.
[142] Onfray, Michel: *La raison gourmande. Philosophie du goût*, p. 70.
[143] Grimod de la Reynière, Alexandre-Balthazar-Laurent: *Almanach des gourmands. Huitième annèe (1812)*, p. 9.

In 1530 Erasmus of Rotterdam (1466–1536) published *De civilitate morum puerilium* (*On Civility in Children*), a book about manners that was so popular that it was reprinted no fewer than 130 times.[144] However, his ideas of good table manners were far from those we subscribe to today: it was important, he said, that the upper class only used three fingers and not the whole hand when they ate from the common bowl, that there must be no mucus in or under the nostrils, that the clean knife and cup were placed on the left and the bread on the right. Forks were still not common in his day, nor indeed were platters; for the most part there was only bread on which to place the meat and any vegetables, and more often than not the cup was shared.[145] One brought one's own knife along. The meat was usually placed on the table as a whole roast, from which each cut their own portion. The fork did not come into common use until the beginning of the 17th century among the wealthiest, and even then it was only employed to take from the dish. Not until the 19th century did the fork and its prongs replace the hand. But that sort of change also affected the relationship between people in general. Elias writes:

> People who ate together in the way customary in [the] Middle Ages, taking meat with their fingers from the same dish, wine from the same goblet, soup from the same pot or the same plate . . . such people stood in a different relationship to one another than we do. And this involves not only the level of clear, rational consciousness; their emotional life also had a different structure and character. Their affects were conditioned to forms of relationship and conduct which, by today's standard of conditioning, are embarrassing or at least unattractive.[146]

Table manners and regimentation are of course important for relations between people, but they also affect the relationship to the food. The fork is the prime example of how it becomes good manners to control the direct, bodily contact with the food:

> The fork is nothing other than the embodiment of a specific standard of emotions and a specific level of revulsion.[147]

Until the 16th century it was also common for eating to take place where the food was prepared, but among the nobility the custom gradually arose of separating the kitchen from the dining room.[148] It is a distaste of the natural that manifests itself as a "cultivated distance". The German philosopher Georg Simmel also writes on this:

> Eating with one's fingers has something decidedly more individualistic about it than eating with a knife and fork, since it associates the individual more directly with matter and is the expression of a more unreserved desire. In so

[144] Elias, Norbert: *The Civilizing Process. The History of Manners*, p. 47.
[145] Ibid. pp. 49–50.
[146] Ibid. p. 60.
[147] Ibid. p. 107.
[148] Farb, Peter and George Armelages: *Consuming Passions. The Anthropology of Eating*, p. 199.

> far as the eating utensils moves this desire a certain distance away, a shared form favouring the cohesion of several persons covers the eating procedure in a manner that does not exist at all in the case of eating with one's hands.[149]

Simmel regards cutlery as a sign of the regulating, socializing conditions that are part of the norms of good manners, yet that simultaneously allow for individual liberty. In contrast to the communal eating dish of former times, the modern plate becomes a symbol of individuality; this food is only for me and is not to be shared with others. The plate is the symbol simultaneously for individuality, the norm and the limit.

The gastronomic meal distances itself from nature through more than just its manners. For Grimod it was significant that gastronomy released food from the tyranny of the seasons. The ties of nature involved an irregular supply – so that what one wanted was not available – and a monotony due to the dependency on limited local produce. Both of these necessities gradually changed from the 18th century onwards. The developing transport system and the cultivation of new exotic vegetables increased the stability and variation of supply. The culinary art was set free.

In its physical sense gastronomy is the preparation for taking the Other into oneself, and is a process to which much time and attention is given. The natural origin of the food should no longer be apparent on the table; nor should the food be experienced in its extreme form or its otherness. Via cookery, which is the extracorporal expansion of digestion, the otherness is already more or less digested: we do not experience the finished dish as otherness, for it has already undergone the first part of digestion in the kitchen. The otherness in the dish is already in the process of the transformation that will enable the otherness to be incorporated in us. Cookery makes the inedible edible. When the food appears, the otherness is already being broken down and is no longer so radically "other" that one cannot imagine it as oneself.

3. THE POWERLESSNESS OF THE POLITICAL CONSUMER

So far we have seen that the rationalization of agricultural practices brought about an increase in food production, which not only supplied the local workers and families but also allowed an export from the farms to the cities. This rationalization was linked to the increasing importance of the market and the demand for food, a demand that was due to the birth of the consumer.

In his recent book *Histoire des choses banales*, Daniel Roche (b. 1935) has analysed the genesis of consumption in France. In a sober study drawing on historical documents and statistics he shows that by the end of the 17th century it is possible to trace an increasing consumption of products, not only among the

[149] Simmel, Georg: "Sociology of the Meal", p. 132.

aristocracy and the richer merchant class but also among the lower classes of society. This is also pictured in another way, as soon afterwards the first economic analyses of the new and growing market saw the light, such as Richard Cantillon's *Essai sur la nature de commerce en général* (1755).[150] At the birth of consumption – in France in the 17th century – it was especially attached to food; on average between half and two-thirds of all incomes were spent on food.[151]

In a biological sense man has always been a consumer, but in the modern mercantile sense the concept of the *consumer* is that of one who consumes commodities produced by others. This may be obvious to modern man, who knows of no other kind of consumption, but in the past the change from local self-sufficiency to consumerism was far-reaching, even though it only developed gradually over centuries. Its consequences are seen not only in the new economics but also within ethics, which can no longer be limited to the short range but must follow the path of commerce and thus extend in time and space. We shall look more thoroughly at these ethical consequences in Chapter Four.

Biologically, man is classified among the heterotrophic organisms. These are consumers of organic matter, which has been produced by autotrophic organisms, also called producers. Autotrophic organisms like plants use inorganic substances such as CO_2, NO_3, NH_x, Ca and so on, and using the sun's energy they synthesize composed organic matter. Consumers cannot therefore exist independently of the producers. In agriculture, the production of the autotrophic organisms is suited to the needs of one specific consumer, the human being.

Food is the substance of the metabolism that keeps the households alive, both of the individual and of society. Food production and food consumption are the metabolism of the individual and society; here we find the flow of nutrients that has its origin in the fields and then passes into the food industry, the stores, the kitchens and onto the dining tables. Here it is eaten and digested, and what becomes of it in the end is a substance that is not always easy to handle. From an ecological point of view it should be recirculated, i.e. be brought back to the beginning, to the fields, and thus make up a closed circuit. This is the flow of nutrients and energy that streams through society and sustains life.

Metabolism characterizes societies. Food culture includes not only the habits of eating but also the shaping of landscapes and nature by agricultural practices, the structuring of working conditions in food-producing companies, the handling of faeces and so forth. Eating habits have consequences not only for the consumers but also for large areas of society. However, because production, distribution and consumption of food occupy central positions in culture, food is also about *power*.

[150] Roche, Daniel: *Histoire des choses banales. Naissance de la consommation XVIIe–XIXe siècles*, p. 27.
[151] Ibid. p. 243.

A dislocation of the power balance on the food market has appeared recently with the emergence of the concept of the political consumer. Previously, the owners of production machinery were in charge of all decisions on the practices of production, but the emergence of the concept of the political consumer suggests that the consumers have gained some influence on production practices. Here I understand power as a category that describes the ability to do or effect something, or to act upon a person or a thing. Power is the ability to make a difference, and the relation between ability and difference is understood in terms of *action*. The concept of the political consumer indicates that a new power has been installed, that the ability to make a difference to production practices comes from a new part of society. Political consumers not only consume commodities; their vision is rather that consumption should be realized as politics. Consumers not only consume food, they consume politics too; so their taste for ethics is creating a new food policy.

At least that is how it seems. The political consumer is often described in terms of an *actor*, i.e. as a subject with specific ethical and political interests. The consumers' power is assumed to lie in their ability to dominate producers through shopping practices. The display of power is considered to be a mechanism through which the consumers, via the selective spending of money, exert specific influence on the producers, who must be obedient to their demands. Otherwise their balance will be negative and their production threatened. Food companies must be careful to keep their "ethical profile" at a decent level for fear that the political consumers will undermine them. Ethics has become a sales and competition parameter for them.

Voting with money is considered to be a new form of democracy. All citizens can participate any time they spend money – except the poor, of course, who are excluded because they have too little of it. The politics of the consumers are based on a difficult balancing of the satisfaction of one's own needs and desires on the one hand with the ethical regard for nature and for others involved in, or influenced by, production practices on the other.

Thinking of the political consumer as an actor is the *substantialization* of the concept. The concept of power is understood in an almost mechanistic way, as the master who states the law that has to be obeyed. Foucault writes:

> [C]onfronted by a power that is law, the subject who is constituted as subject – who is "subjected" – is he who obeys.[152]

This simple understanding of the concept of power, which has dominated the theory of power in political science for many years, is attractive because it indicates a cause and an effect: power is understood as the ability to dominate others. Foucault calls this understanding of the concept "juridico-discursive",[153]

[152] Foucault, Michel: *The History of Sexuality*. Vol 1, p. 85. *sub* means "under" and *ject* means "thrown".
[153] Ibid. p. 82.

because it is based on jurisprudence: laws, rules, regulations, directives, censorship, etc., and discursive, in that laws in human society are formulated in words, most often as prohibitions. According to Foucault, this power only has the force of the negative on its side: the power to say no and to post limits. In the last part of Chapter One we saw how, on the present food market where consumers have almost no knowledge about food production practices, a consumer food ethic would only be possible in the form of sanctions – the power to say no to certain products. Despite this fact and its agreement with the "juridico-discursive" view of power, the concept is inadequate for the analysis of the political consumer for the following reasons.

1. **Lack of knowledge and transparency.** The idea of the political consumer rests on the assumption of a transparent market in which the consumers have sufficient knowledge to make informed choices and thus act by shopping in agreement with their ethical and political points of view. However, the claim to knowledge is by no means met under present market conditions even though there are efforts to increase information by labelling and declarations.[154] This is a paradox: the amount of information on food is extensive and already far too much for most consumers, but there is still not enough information to make well-founded political or ethical decisions in the shop itself. The political consumer has become an overloaded information project in the shape of food declarations, public relations (PR) campaigns, media news, consumer organizations, databases, the Internet and so forth. The amount of information is so great that it defeats its own purpose. The practices of food production are not incomprehensible in themselves, but the obscurity of the food market and the deliberate misrepresentations of facts in, for instance, publicities are due to the density of the many transparent layers on top of each other.[155] To begin with, there are layers of different activities: the breeding of plants and animals, the industrialization of agriculture, the technological development in food-processing practices and the complexity of the logistics of food distribution systems. Well-intentioned information campaigns can do very little about this, as the problem is not the understanding of the information but the very amount of information. Labels that are intended to make it easier for the consumer only render it more complex because there are so many of them that it is often not clear what they stand for and because new labels are constantly emerging as old labels disappear. The knowledge required to understand the scientific, technological

[154] There are, for instance, approximately 500 different labels describing specific qualities of products in Denmark. *Mærkningsudvalgets redegørelse*, report from The Danish Ministry of Trade and Business, Denmark, 1999.

[155] See Lévinas's short article, "La trace de l'Autre".

and logistic implications of food production practices in order to make an informed choice is highly advanced and cuts across sectors. Who among consumers knows, for instance, that it is common practice to remove the sprouts of the grain seeds before they are milled to flour? Or that most of the vitamins and also the flavours are actually to be found in the sprouts? The scientific, industrial and technological development of food practices have occurred almost without the knowledge of the consumers; they hold on to the romantic idea of food production from the good old days when a carrot was a carrot and flour was made of whole grains and contained no additives. But those days are long gone. The conclusion is that as long as consumers have no knowledge they are unable to make a difference.

2. **The consumers' lack of autonomy.** As we have seen, the exertion of power on the food market is located in the individual consumer as its substantialization. In her reflections on the nature of power the German-American philosopher Hannah Arendt (1906–1975) writes:

   > Power corresponds to the human ability not just to act but to act in concert. Power is never the property of an individual; it belongs to a group and remains in existence only as long as the group keeps together. When we say of somebody that he is "in power" we actually refer to his being empowered by a certain number of people to act in their name.[156]

   In Arendt's view, the individual consumer is more or less powerless because power is exerted not by the individual but by a group of individuals. The relevant question here is: what makes consumers act as a group? And how can consumer groups (or segments) be influenced by others to make specific choices or avoid certain products? Among the answers are: information campaigns, a good press, a lack of bad press, PR campaigns, labels, price policy and so forth. Indeed, it looks as if the autonomy or self-determinism of the individual political consumer is an illusion, partly because there is not enough information to make an informed political or ethical choice and partly because the consumers' choice is well prepared by others in advance. Too much is at stake to let consumers decide of their own free will.

3. **Limited options.** There are no other options than the ones offered in the stores by the companies. When so few companies dominate the market, there is often only one brand available, and the consumers' "choice" is to buy this brand or none at all. This is another paradox of the present market. Even though the whole chain from "farm to fork" appears obscure to the consumers, shopping is amazingly simple: most of the time there are only very few products to choose from.

[156] Arendt, Hannah: *On Violence*, p. 143.

The substantialization of power is a tempting concept, in spite of its obvious problems; and there are several reasons for this. First, consumption is linked to the self-understanding and identity of the consumers.[157] Consumption gives consumers a feeling of being able to make a difference. It would be foolish to state that the consumers make no difference in preferring one product to another. I would not argue for this view. Rather, my claim is that the lack of knowledge makes the consumers unconscious of the consequences of their choices and that they might make a different choice if they had better knowledge of the production practices behind the products. Or, in the most radical case, they might completely stop consuming certain kinds of products. Second, causal explanations are preferred because they give a simple understanding of how things work. If we give up explaining, room is created for the unknown. The unknown creates uncertainty and maybe fear, whereas the known is reassuring. That is at least one good reason for replacing the unknown with the known.

If the substantialized concept of power is of limited use in the understanding of the political consumer, how are we then to understand power? Nietzsche was less interested in power as the ability to make a difference; instead he describes power as the deployment of regularity and repetition. Power is conservative in the sense that it seeks to conserve things as they already are and to exclude things that are not yet, but that *could*, come into being. It is a kind of control, which seeks to establish perfect predictability and exclude the unpredictable, as the latter could present a threat to those in power. Thus, the kind of predictability installed by power is not the one used to make reasoned decisions, but the one to conserve already existing power structures.

Reflecting on the political consumer in the light of Nietzsche's pessimistic philosophy gives rise to a nasty suspicion: the concept of the political consumer says nothing about the consumers' *actual* power, since their lack of knowledge takes the edge off their ability to make a difference. Rather, it is a concept used to construct a new reality in the minds of the consumers. I shall try to follow this thought to the end, despite the risk of ending up with some fictitious and paranoid conspiracy theory. For even though it might not be the whole truth, it might be a part of it. Following the concepts of power in Nietzsche and Foucault, the idea is that of course the concept of the political consumer makes a difference; it is just that it is not the consumers themselves, but rather the articulation of the concept, that makes an actual difference – merely by its very existence.

It would be imprecise to talk about a definition of power in the works of Foucault; rather, he seeks an understanding of the complex nature of power and the many ways in which it is practised. In Foucault's view power is ubiquitous because it emanates from everywhere. Understanding power as the relative

[157] See Chapter Five.

strength between actors accentuates its dynamic and relative dimension, as in the unstable relation between two or more positions. The accentuation of the strategic dimension emphasizes the many different ways in which power is practised as well as the many guises in which it appears. Mostly, power is not directly visible.[158] The display of force is lost in a network of actors and their interests, in relations between people and their organizations, where it might be possible to see who makes the decisions but not the influence the many different actors might have exerted. The display of power even holds an interest in being secret, as it makes it more difficult to exert resistance to it.

It would be interesting to analyse the genealogy of the concept of *the political consumer* and to discover the origins of the phrase to see the original intentions behind the concept. However, it is far more important to understand how it is used today, how it influences our thoughts and thereby constructs new "realities". For, as I have shown above, the political consumer is not just an actor on the market. *The political consumer is a concept used in the strategic display of power*. Foucault's insight lets us see that the emergence of the concept of the political consumer is the articulation or the historical deployment of a specific issue with a strategic purpose. To give something a name is in itself an act that makes a difference. It is not just the simple act of putting a word to something that everybody already knows. Putting a name on something makes it possible to direct the public attention and discourse. It makes a difference by extending the agenda for what can be discussed. In the words of the Danish writer N.F.S. Grundtvig (1783–1872): "[T]he word creates what it names".

Now, if the concept of *the political consumer* is used strategically, we might assume it would be in order to legitimize existing power relations, which in this case are those of the food market and its current stakeholders. When consumption is understood as one of the most efficient ways of making a difference, we are indeed tempted to see and understand ourselves as consumers. The articulation of the political consumer is a strategically very subtle display of power that never questions consumption in itself. Consumption becomes good per se as long as the ethical profile of the manufacturers is good. In praising the political consumption it is, however, often forgotten that the political aspects of consumption are mostly illusions due to the lack of information and transparency. In reality, this praise of political or ethical consumption thus tends to be praise of the liberal market as it is.

If the word *power* is to have any meaning, it must of necessity be accompanied by some kind of resistance to the display of that power. Otherwise the power would not make any difference and thus not be any power at all. For this reason, a disregard of, or indifference to, the display of power – which can be seen as one way to avoid any resistance while not actually obeying – is often

[158] Lukes, Steven: *Power: A Radical View*, p. 24.

said to delegitimize power. The irony and tragedy of political consumers is that they believe themselves capable of exerting resistance through political consumption. But this resistance, where consumption is moved from one kind of commodity to another, just serves to conserve or even increase consumption and thus to expand the market.

This theory (which I have already delineated at length and which leaves me with a sense of powerlessness) suggests that those consumers who see themselves as political do not merely tolerate the power of the market but even appreciate the staging of themselves as powerful actors; and in so doing, of course, they actually contribute to the illusion and the secretiveness of the display of power. For there is no reason for consumers to give up that illusion, as it is important for the continued satisfaction of their needs for commodities. This is the unspoken alliance between consumers and producers, where the real losers are those who cannot afford to pay: those without money.

The political consumer is *not* an actor on the market. It is a concept used for the strategic display of power on the market, available as a threat or to exert pressure or as a tool in the struggle for market shares, where ethical concerns about the environment, health, working conditions and so on are precisely calculated. Creating an ethical profile for the company is a new strategy for selling products.

This description, based on the works of Nietzsche and Foucault, deliberately gives a negative picture of the use of ethics and the concept of the political consumer on the market. This does not exclude it from being used in a more sober way, however. Seeing the concept of the political consumer only in the light of a conspiracy theory is too pessimistic. The articulation of the concept and the debate surrounding the political consumer have undeniably had some positive influence on production practices by forcing some companies to be more concerned about ethical issues, even if it is only in order to increase competitiveness. The willingness of the producers to change production practices according to ethical standpoints, even though this usually increases expenses, is evidence of the power of the political consumer. Nothing is changed in the dominant economic order, but certain practices of production have been altered. To put it in economic terms, one could say that some companies have internalized some externalities: production costs that were formerly ignored by the companies (like the disposal of waste water) are now in some areas increasingly part of their domain. And of course ethics is not only about costs in an economic sense; loss of animal welfare, for instance, can never be refunded.

Consumers with a taste for ethics might well think of themselves as political consumers, but it is nevertheless important to be aware of the ambiguity of the concept. It is true that political consumers can make a difference, but strategically the concept can be used just as well by others, e.g. producers, in order to make a difference. As a political consumer one might ask oneself: what difference does my consumption imply for nature and society? This is seldom easy to answer.

4. THE HIDDEN PRODUCTION HISTORY OF FOOD

Throughout history man has practised different ways of gathering food. The first *Homo sapiens* were hunters and gatherers. They took what nature offered of edible animals and plants. Food resources were scarce and huge areas were required even for smallish groups of human beings. The first domestications of plants were a true revolution, as it now became possible to supply larger amounts of food from smaller pieces of land. This increased the density of the populations. It also meant that people settled on the soil and gave up the lifestyle of the nomad following the migrations of the herd.

In the ancient world the most commonly used word for cereals (such as the ancestors of wheat: emmer and einkorn, but also barley) was *sitos*.[159] The first agrarian cultures emerged about 10,000–15,000 years ago in the Mesolithic period, when human beings started to cultivate the fertile lands of ancient Mesopotamia. Farming made it possible for human beings to settle, to stay at one *site*. The residents were soon to be opposed to the nomads, who like shepherds followed the herds, i.e. the food. The Amorite belongs to the nomadic people:

> The Amorite who knows not grain . . . . A people whose onslaught is like a hurricane . . . . A people who have never known a city.[160]

The feud between the farmer and the nomad is illustrated by the dispute between Cain and Abel of the Old Testament. The sons of Adam inherited an equal division of the world: Cain the ownership of the land, Abel of all living creatures. Abel is a nomad and breeds sheep. However, Abel is more civilized than the early nomads as he actually has domesticated the sheep. Cain cultivates the land and is a settled farmer. In comparison to the free life of Abel, the life of Cain is dominated by the hard manual work of digging and tilling the soil. Cain envies the free life of Abel and kills him. For this Cain is sentenced to wander in the land of Nod (the land of wilderness or desert).

Cain, the farmer, commits a murder and is sentenced to a reconciliation wandering, but in the long run the resident farmer becomes the more successful of the two. One can only guess as to why the development went this way, but one essential element is that the cultivation of the land increases the number of people who can be nourished on a given piece of land compared to hunter-gatherer

[159] Isager, Sigen and Jens Erik Skydsgaard: *Ancient Greek Agriculture*, p. 21.
[160] Sumarian text quoted by Bruce Chatwin in *The Songlines*, p. 192.

cultures. When the number of people increases, it is not hard to imagine that at a certain time new kinds of settlement appear – at first as small towns and later on as larger cities. The cultivation of land and the domestication of animals are the basis for the appearance of towns, which by definition nomads cannot build. The emergence of towns changes the structural basis of societies. It is characteristic of towns that they have some kind of division of labour. A town needs supplies of food, but not all the inhabitants are involved in the cultivation of the land. The emergence of towns is also the beginning of a more systematic division of labour into specific professions. Although hardly visible at the outset, the division of labour, which is so familiar to us today, has its origin in the cultivation of the land and the emergence of towns.

In industrialized countries only a very small percentage of the population is occupied with the tillage of the soil. Food is a product that consumers buy – and about which most of them know very little. Food processing and distribution is a complicated procedure, which involves many actors and this obscures the origin and the history of the food. The link from the kitchen back to the story behind the food stops at the shelves of the stores: most distribution systems do not allow for the consumers to relate further to the history and the origin of the food. The information on the declaration of contents is all there is. Moreover, the preparation of food is increasingly taking place outside the household,[161] a fact that increases its obscurity even more. Convenience food must be made easy for the consumers; not only is its preparation minimized but it is a general characteristic that it also reduces the efforts of the consumers to a minimum. Leave the inconvenient things to the producers – that is the motto of the producers of convenience food. And if the production history is unpleasant, it is convenient not to know anything about it.

The distance between the kitchen and the cooking on the one hand and the origin of the food on the other is growing. The preparation of food could actually designate the whole process of making food – from farm to household – but we use the term in a much narrower sense, i.e. as cooking. Cooking starts as a reflection on what to eat and how to cook it. This reflection typically takes place before or during shopping, or more spontaneously as an "improvised course" based on the food available in the kitchen. The cultivated distancing to the origins of the food continues to grow; it no longer comes into the kitchen directly from the fields, from staples or from nature. This is even reflected in the design of the kitchen (if there is a kitchen at all!), where the scullery has vanished. Modern kitchens have no place for cleaning newly harvested vegetables or slaughtered animals. There is no room for soil, dirt, blood, bones or entrails. Modern kitchens are constructed to deal with clean and highly processed foodstuffs.

[161] Meyer, Rolf and Arnold Sauter: "Bedürfnis-und Konfliktfeld Ernährung: Handlungsfelder für die Politik", pp. 29–32.

There are at present two tendencies in the development of household kitchens that at first sight seem contradictory. One is to abolish kitchens completely and thus give up cooking at home; instead of eating privately at home many prefer to eat in public. The other tendency is in the opposite direction and makes the kitchen the centre of the house, the common place, where the family members gather around common activities. However different these two tendencies seem, they have the same background: the cultural distancing to the natural origin of food. Clearly, a disinterest in cooking for oneself must create a distance to the food, yet the modern focus on kitchen design would probably not have been possible without a food-processing industry, which transforms the raw and even repulsive-looking food into a highly cultivated product. What comes into the kitchen is not dead animals with skin, hair, eyes and entrails, or carrots with green leaves and soil on them. The food that comes into the kitchen today is less natural and more cultivated, and that is why we do not need to create the same cultural distance to the kitchen as we did in earlier times. Home cooking is no longer about transforming nature to culture, since what comes into the kitchen already belongs to the sphere of culture. This is how the kitchen has regained a central household position in modern society.

There is something abstract about modern kitchens. As less and less emphasis is laid on the production history of food and its cultural embedment, cooking seems almost metaphysical. The kitchen and the practices of cooking have become emancipated from temporality and space. Cooking has lost its relation to seasons and spaces. The seasons have virtually no influence on the supply of foods and there are no geographical restrictions when it comes to the transportation of food. Time and space are suspended in the kitchen. In the modern kitchen, all the seasons coexist simultaneously and one can move freely from one season to another. Or they can be mixed *ad libitum*. Fresh apples and strawberries carrying the flavours of summer, lightness, deliciousness and green landscapes can be enjoyed independently of the season, even on the coldest days of winter. For the modern kitchen, the season of the harvest continues all through the year. And there is never any shortage. There are no seasons when people suffer privations. There is always some place on earth where new potatoes can be harvested and transported to our kitchens.

The supermarket is the window of the kitchen that looks out towards the world. The supermarket is a miniature representation of the world, although the focus is not centred on geography but on the culinary. Represented on this culinary globe are all the climatic zones – tropical, subtropical, temperate and polar – with all the edible plants and animals living on the world's seven continents: Africa, North America, South America, Europe, Asia, Australasia and Antarctica. In retaliation for the lack of seasonal variation, supermarkets have introduced their own seasons: the January sale, the summer sale, etc. These are the new *artificial* seasons, which have more relevance to cooking than do the climatic seasons.

The emancipation of the kitchen from time and space implies a paradox. On the one hand, the disengagement from local space and temporality constitutes a release from the necessities and the limits set by space and time. On the other hand, this emancipation is simultaneously a rupture with a more local and corporal relation to food. At present, globalization tends to reduce local production diversity to a minimum because of the benefits of large-scale production. However, consumers are typically much more familiar with local food than they are with "globalized" food. The history and the direct impact of local food production practices are typically part of the consumer's everyday life: agriculture forms the landscape, and food industries and food stores influence many areas of the cities. There is a temporal, historical and spatial knowledge connected to seasonal and local food production practices.

The emancipation of the kitchen in time and space entails both disengagement and isolation. In one and the same movement, where the kitchen is emancipated from the tyranny of space and season, and where there is a rupture from temporal and spatial isolation, another isolation is created. This is the isolation from the local context, from local time and space.

Even though this emancipation of the kitchen and cooking is an opening towards the world, it also articulates the modern bluntness of the kitchen and the consumers. In the blunted kitchen, all the edible things from all over the world are available, but as open as its horizon seems to be, it is nevertheless very limited. Since we know almost nothing about the foodstuffs, all that remains is the physiological taste of it – the pure aesthetic experience.

Cooking in itself has also become an activity that intends to hide the origin of the food and its production history. Cooking has already been described as an art of transformation, where nature is transformed into culture. Cooking is more than preparation; it is a ritual that makes the inedible edible. In more general terms, it is the transfer from one category to another. Mary Douglas describes the meaning of the ritual as a

> focusing mechanism, a method of mnemonics and a control of experience . . . which shut in desired themes and/or shut out intruding ones.[162]

A ritual makes a frame, says Douglas. The rituals of cooking and table manners are frames that carefully prevent us from considering food in its natural origin and making us focus on its cultural attributes. For most foodstuffs the metamorphosis is essential; we do not want to be reminded of the organic origin of the food – of the life that it once was, that bacon was once pig, and for this reason we focus our attention on taste and aesthetics. Our attention is directed towards desirable themes and away from other themes. The rituals of cooking and eating are means of controlling this experience. The preparation and aesthetics of the table ensure that we do not experience the beef as a cow, but keep

[162] Douglas, Mary: *Purity and Danger*, p. 64.

the beef in the regime of the edible. The aim of the art of cooking is to keep us in the right mood, i.e. as cultivated eaters who do not eat the other, nature, but the same, the cultivated or culture.

As consumers of food, we focus on food in its aesthetic form and shut out its ethical perspectives. It is no longer obvious that food has an origin and a production history. To most consumers, agriculture, the food industry and the other enterprises involved in bringing food to the food stores and the dinner tables are, to all intents and purposes, hermetic areas that they know precious little about. Consumers are unable to look back on the food's production history, and consequently they are equally unable to see how their own food consumption influences nature and society. The relations are lost. The coherence between the different parts in the chain from "farm to fork" that comprises the system of food production is concealed from consumers.

Instead of viewing food as something that has a production history, the tendency is to regard food as something that can be clearly defined by food declarations. This, however, happens at great cost. Nietzsche writes:

> [A]ll concepts escape definition that summarise semiotically an entire process; only that which has no history is definable.[163]

When a concept has a history, its meaning is likely to have changed during that history. In the terminology of Koselleck this is called *Begriffsgeschichte* (the concept-history). This also implies that the concept has several meanings, which in turn can make it ambiguous. The word "food" is of course a concept and it too has its own concept-history. However, the interesting point here is not so much the concept-history of the word as the way that we describe and define food through food declarations. Food declarations are substantial and materialistic as they define the ingredients of the food and thus necessarily exclude intruding production histories. These stories cannot be defined; they can only be experienced or retold.

5. DO NOT EAT WHAT YOU HAVE NOT READ

We have acquired the habit of *reading* the labels and declarations on the food if we want to know something about it. This is a new movement; one holds the food in the right position and at the right distance to the eyes so that one can read about its content. Often this is the only available knowledge about the food. The employees in the supermarkets can usually only tell you two things about the requested article: its location in the shop and whether it is available at present. But this is knowledge not about the food, but about the shop.

[163] Quoted from Koselleck, Reinhart: *Futures Past. On the Semantics of Historical Time*, p. 84.

Labels represent a discursive and standardized procedure of recognition, which admittedly makes shopping faster and smoother but simultaneously excludes most sensuous perceptions. Those who find this to be an exaggeration should try forbidding themselves from reading the labels next time they do the shopping. This is of course the permanent situation of illiterate and many dyslexic people, for whom bringing home unwanted products is a recurrent problem. Dyslexic people have difficulty in distinguishing between sugar and flour, for instance, because packaging is very similar and the content well hidden. If one cannot read the labels, there are many products that simply cannot be distinguished without previous knowledge of the colours, brands, logos, etc. of the products. But if one is forced to open the packing and wrapping, one can see, smell or taste the product.

Food labels are placed on the packing. The packing and wrapping hide the food and prevent consumers from sensing it; only to the eye does part of the food remain visible. In its packing or wrapping it excludes the consumer and also wraps the food on a physiological and psychological level. Food labels reduce sensuality and corporality "mentally" by appealing to the reason of the consumers. The intellectual reflection, the concentration on the food declaration, limits sensuality. And of course vice versa: intense corporal and sensuous activity also exclude reasoning and reflection to a certain extent.

The intellectual focusing on food declarations and labels leads in most cases to the reduction of sensuous perceptions because there is nothing that triggers or directs the senses. Thus food declarations and labels that are intended to inform the consumers often have the opposite effect; they hide what they wish to show.

Food declarations are expressions of rationalized food production practices, the food-processing industries and the rendering of food into a scientific object by food science. Simultaneously, they are a signal of the inadequacy of the senses, of their inability to provide us with the right information on food. Food science gives us information about the food that the senses could never provide. Food declarations teach us that the insight provided by science is more adequate than mere awareness through the senses. For the senses are not objective; rather they are subjective and subject to personal taste. The sensuous and bodily propensity to be attracted to what is forbidden and unhealthy further disavows the senses, not only as unintended misinterpretation but also as intended and harmful delusion. We all know (maybe because we are taught so?) that if we let taste and desire prevail, it leads to an unhealthy diet.

However, within biology it is often assumed that human and all other living beings have developed their senses in areas of special importance to them and according to their needs. The senses must provide information about the environment to the organism in order to enable it to react and "navigate" in the environment. Certain things are perceived while others are not. For instance, different organisms are sensitive to different kinds of light. It is of vital importance that organisms are able to orientate themselves according to food preferences. The senses must be able to provide information about what is edible and what is not. Organisms that do not have that capacity will not survive long. It is often stated that human beings differ from other animals in that, given various alternatives, we are able to make free choices. With animals, however, it is controversial as to what extent they are able to make "free" choices or whether they merely follow their instincts. Phytognomy, an old and disregarded science from the 14th century, was based on the assumption that in nature every poison has an antidote. On this premise, when an animal is sick, an antidote can be found in nature – in the plant kingdom. Specific plants have a curative effect on specific diseases. The instincts of animals are thought to provide them with information about these "cures", enabling them to cure themselves at least partially (depending on the disease of course) by eating selected plants. It has been observed that when cows are given free access to meadows with high biodiversity, they are very discriminating about what they eat, and some even argue that they are able to cure their own diseases – to a certain extent of course. Is there in human beings a residue of instincts able to guide us about the effects of plants on our bodily and mental health? It is hard to tell, but in any case we do not trust our perceptions in this respect. And with good reason, I must add. During the evolution of man, or more precisely the development of man from a natural being to a cultural being, the ability to reason seems to have been won at the cost of instincts. Evolution has certainly not come to an end; the modern intellectualization of food continues to displace sensuality.

Perhaps this is not so bad after all. If most food in the future is going to be based on GMOs, where genes can be transferred from one species to another, senses and instincts are not going to be of much use anyway as a source of information about food. We should then get used to thinking of a carrot not as a carrot as such but as a carrot with certain genetically modified qualities. There might be genes and proteins from bacteria, fish, animals or even human beings in our carrot. The senses and instincts will probably not be of any use in that case; instead we shall have to trust the food declaration.

In its origin, the cultivated distancing to food and eating was brought about by simple technologies such as fire and knives. Later on also the use of plates and forks contributed to this development. Today, this cultivated distancing has extended its scope through modern food production practices: food science, industrialized food-processing techniques, gene technology, standards for food hygiene, etc. The cultivated distancing is the dissociation from what is traditionally considered the opposition of culture: nature all around us.

Nature is not eaten raw. Most food stems from nature, though it is a very cultivated nature; animals are domesticated, plants are bred, raw foodstuffs are washed, processed and wrapped before eating. We avoid sensing the food in its natural state. In spite of the trendy rural kitchens and the focus on authenticity, the fact for the modern consumer is that the aesthetics of the food are reserved for the transformed and processed. Our senses are closed to the natural and opened to the transformed – to the cultivated meal. Taste and sensuality are reserved for the cooked food and cultivated meal, and what is placed before the dinner to that we close our sensuous perceptions. Food is nature, which must not be perceived as natural.

This is not to say that the modern kitchen and meal are without sensuality. On the contrary, they seem more aesthetic and sensuous than ever before. We are indeed very concerned with the taste and aesthetics of food. This desire for sensuality and aesthetics around the dinner table may be seen as a consequence of the limitation of sensuality to the transformed foodstuff. With our sensuality to the meal and the processed food now reduced, the meal must compensate for the loss of sensuality that was formerly associated with handling food before dinner, i.e. the natural aspect of food. The traces of authenticity and closeness that are related to the food in its raw condition must now be provided by the cultivated meal. The conclusion is that the meal has become more aesthetic than ever, but also that the foodstuffs we buy have become more intellectual than ever.

We eat information. Consumers' knowledge about food is in most cases reduced to what can be read on food declarations. Thinking about how the food is produced we tend to revert to romantic and old-fashioned illusions: milk comes from cows and bread is made from flour. That certainly is true, but we do not pay much attention to the processing practices of modern food production, mostly because our knowledge is very limited. We cannot think about what is unknown to us. Traditional knowledge about food and its production practices is largely outdated. And this schism between the knowledge of consumers and the actual food production practices may become even deeper. For even the sensuous perceptions may be questioned. Gene technology might lead to a rupture between sensuous perception and reality, because the carrot is not only a carrot but it also carries genes from other organisms.

Unable to trust our senses, we rely on what we can read: the food declarations. In future this will become even more important for us: do not eat what you have not read!

PART III

FOOD ETHICS AND THE PRODUCTION HISTORY

CHAPTER FOUR

TRACING THE PRODUCTION HISTORY

Consumers with a taste for ethics want to know about the food production history. Such knowledge can be obtained in different ways. According to the sources it originates from knowledge can be categorized as intellectual, sensuous, emotional and narrative.[164] Intellectual knowledge arises from the use of our intellectual capacity to make rational arguments, to reason and to make judgements. Part II of this book is a criticism of the dominance of this kind of knowledge in our relationship to food. Part III, in contrast, is mostly concerned with the other sources of knowledge. Sensuous knowledge is acquired from the use of the senses, in the present case from the phenomenological interface between food and our senses. Some even prefer the phrase corporal knowledge. Emotional knowledge is the knowledge that can be felt inwardly in our emotional engagement in others and in nature. It is knowledge that need not be clearly articulated as thoughts (but is known by feelings). Finally, narrative knowledge is knowledge acquired from a story, either lived or told. Narrative understanding takes up a central position within the hermeneutic tradition.

However, Part III is not only about epistemology and knowledge. I shall also return to the question raised in Chapter One concerning the link between the food we eat on the one hand and the ethics involved on the other. We saw that such a link must exist, since we know that foods necessarily have a production history and that this story can be everything from good to bad to evil. First of all we should notice that food is seldom conceived as pure substance – Feuerbach's view being a provocative exception. Food is not just another kind of fuel; it cannot be compared to the gasoline we use in cars. Food takes up a central position in our lives; it is not only indispensable for our existence but also central to our understanding of what constitutes a community or society on a basic level. Moreover food is natural in its origin and thus food culture reflects a relationship to nature.

It was also shown in Chapter Three that the information labels on food are of limited use in relation to food ethics. We need therefore to reflect on what it takes for consumers to display their food ethics in real life. I believe two things

[164] See Fernández-Armesto, Felipe: *Truth. A History and a Guide for the Perplexed.*

are essential to give food ethics a real content and empower consumers to act (read "shop") in conformity with their ethical views.

First, I shall point to the importance of sensuality and corporality in relation to food from an ethical and existential perspective. Sensuality and corporality depend on self-experience and are therefore important in bringing food and food production closer to consumers. As will be shown, closeness is an important feature for the general ethical commitment and dedication. Thus, we must examine if the food itself, in its phenomenological appearance, can represent or function as a symbol of its own production history.

Second, I shall argue that by viewing the process of production as the production "hi-story", the ethical implications will be made clearer to consumers. I consider the narrative form essential for the application of ethical judgements, and I shall for this purpose use a special kind of semiotics, that of thc *trace*. I shall examine how the food itself can function as a trace of its own production history.

The question is, how can the perception of the food itself or something relating to the food lead to the narrative of the production history? The use of food in the Christian Eucharist meals shows us how it can be regarded as a medium to inspire meditation on God. This is very useful; since we have to eat every day (except during fasting), Christians, especially those who say a prayer at table, are inclined to think about God several times a day. Rather than meditate on God, however, I am here examining the ethical reflection brought about by the food we eat: how food can be seen as a "medium" for our relationship to others and to our surroundings.

1. SHORT-RANGE AND LONG-RANGE ETHICS

In his book *The Imperative of Responsibility* (1979), Hans Jonas describes the way in which our increasing powers of technical intervention in nature are influencing ethics. The influence is considerable, because the consequences of intervention in nature, and in human nature, cannot be limited to local matters but need to be understood in a wider spatial and temporal context. Of the ethics of antiquity Jonas says:

> The good and evil about which action had to care lay close to the act, either in the praxis itself or in its immediate range, and were not matters for remote planning. This proximity of end pertained to times as well as spaces. The effective range of action was small, the time span of foresight, goal-setting, and accountability was short, control of circumstances limited. Proper conduct had its immediate criteria and almost immediate consummation. The long run of consequences beyond was left to chance, fate, or providence. Ethics accordingly was of the here and now, of occasions as they arise between men . . .[165]

[165] Jonas, Hans: *The Imperative of Responsibility*, pp. 4–5.

Human beings had no responsibility for events occurring in the distant future or outside the town (e.g. in other cities or in the countryside). Ethics was connected with personal experience.

Over the years, however, the increasing technological power of man has had an important influence on ethics. This is because our intervention in nature and the human body has consequences that are no longer limited to the local but can be traced over lengthy distances in time and space. Today, the boundary between city and nature and between culture and nature has in many ways been eroded. It is no longer the case that cities are surrounded by wild, virgin nature. There is probably no such thing as virgin nature, even though that is how many people like to think of it. According to Jonas, it is the technological development that makes it necessary to rethink ethics in a broader context than the local and close surroundings. The skill or aptitude in the art of craft defined by the ancient Greek word *techne* was considered ethically neutral partly because it was not thought that technical intervention could cause permanent damage to the world and thereby threaten its integrity, and partly because the thought of manipulating and controlling the world with technology would not have occurred to the minds of the Greeks.

The British philosopher, Alasdair MacIntyre (b. 1912), touches upon the same subject in his book *After Virtue* (1981), in which he too returns to Aristotle, who

> sets himself the task of giving an account of the good which is at once local and particular – located in and partially defined by the characteristics of the *polis* – and yet also cosmic and universal.[166]

This tension between the local and the universal can be found throughout Aristotle's *Nicomachean Ethics*. MacIntyre's critique points to Aristotle's lack of understanding of the transience of the city. He links this to the claim that Aristotle had little or no understanding of historicity in general.[167] MacIntyre also emphasizes the role of technological development in the breakdown of virtue ethics of antiquity and the foundation of ethics in the local, but he does so in a different way from Jonas:

> One of the key moments in the creation of modernity occurs when production moves outside the household. So long as productive work occurs within the structure of households, it is easy and right to understand work as part of the sustaining of the community of the household and of those wider forms of community which the household in turn sustains.[168]

MacIntyre describes how the change in production and the rise of consumption, which in Chapter Three we saw Roche tracing back to 15th century France and

[166] MacIntyre, Alasdair: *After Virtue*, p. 148.

[167] Ibid. p. 159. As shown in Chapter Two the meaning of the word "history" before the 17th century is akin to "description".

[168] MacIntyre, Alasdair: *After Virtue*, p. 227. This is very close to Simmel's sociological definition of modernity, which considers the separation of production and consumption a key element of it.

which is of course linked to technological development, influences ethical thinking. The need for a new foundation and legitimization of ethics is, for MacIntyre, less a result of technological development in itself. It is not, as Jonas claims, man's increasing power to manipulate nature through technology which shows us the vulnerability of the living and in turn makes man acknowledge his wider responsibilities. Rather, it is the lack of closeness and transparency deriving from the separation of production and consumption in the Age of Enlightenment that has led to dissatisfaction with virtue ethics and encouraged many philosophers to legitimize ethics in a new way.

On this matter, MacIntyre is probably right. It is not until much later that the idea of the vulnerability of nature emerges, in such works as Rachel Carlson's *Silent Spring* (1962). However, neither MacIntyre nor the philosophers he analyses in his book are as clear and unambiguous as Jonas in describing the changing spatial and temporal conditions of ethics. MacIntyre concentrates on the dissolution of Greek virtue ethics during the Enlightenment and the ensuing search for new ethical foundations. This is the perspective in which he analyses moral philosophy from the 15th century onwards, including Kant's "categorical imperative" and Bentham's utilitarianism, both of which he sees as new ways of trying to legitimize ethics. This is also why today we are faced with so many different kinds of ethics. Ethics has become a muddle of utility, principles, rights and a variety of virtue concepts. MacIntyre thinks that the philosophers of the Enlightenment failed to give ethics a new foundation. Disagreement about the concept of ethics creates a moral chaos.

The lack of closeness and transparency deriving from the separation of production and consumption is a challenge to ethics because it makes it impossible for citizens and consumers to understand the consequences of production practices, as they are no longer to be found in the local context. This has serious implications for ethics. Ethics can no longer be restricted to immediate and local acts but must be extended in time and space. The Danish philosopher Peter Kemp (b. 1937) has named this phenomenon *long-range ethics* (as opposed to the *short-range ethics* of antiquity).[169] Today it is clear not only that local actions have implications for the near future and the local context, but also that when it comes to praxis, long-range ethics encounters some fundamental difficulties. First of all the concept does not appeal to us as existential and bodily persons, because the consequences of the actions are never made visible and present to us; they never become part of our lives. Judging whether an action is good or evil then becomes impossible. To see if an action leads to the good life that succeeds is no longer possible because we do not *know* what succeeds. However real the consequences are and no matter how concerned we are with acting according to our moral values, the distance in time and space makes

[169] Kemp, Peter: *L'irremplaçable. Une Éthique de la technologie*, p. 99.

long-range ethics seem abstract, fictitious and hypothetical. We are never faced with the consequences of our actions. Hence for Kemp long-range ethics must presume an ethic of closeness, founded on the face-to-face situation:

> To understand the possible consequences of our actions in the long run we have to imagine the temporarily and spatially absent person as a face, as an expression of a living and bodily person we know from the immediate encounter with the other . . . . At long distance we can only act responsibly if we relate to the absent as if we were face to face with them, even if they are people we have never met and never will meet. Our long-term actions are only ethical when they are understood and performed by analogy with our actions in the local and the present.[170]

The idea of acting *as if* the absent were present and, by analogy, in present time is an important step for long-range ethics. But is it enough? We have to ask what it means to act *as if* the absent were present? The complexity of society means that we cannot simply *act* as if the temporarily and spatially absent were present. The complexity obscures the consequences of our actions, and even if we do have a vision of the good life with and for others in fair food production practices, consumers do not know what actions will ensure this good life. The analogy is impossible because actions on the food market take place in temporarily and spatially complex contexts and therefore cannot be compared with face-to-face actions.

Jonas writes that the gap between knowledge and the power of technology poses a new ethical dilemma.[171] The knowledge required for carrying out a good or righteous act is not the same in a real face-to-face situation as in an imagined face-to-face situation. To carry out the righteous action in relation to the absent requires more knowledge. Jonas therefore points out that knowledge plays a far more important role in long-range ethics than in short-range ethics, and that long-range ethics establishes the duty of knowledge. Within long-range ethics knowledge is a precondition for responsible action. Acting as if the absent was present is impossible.

Nor can we simply transfer short-range ethics to the absent. The difference is the knowledge. Kemp is right in arguing that short-range ethics must be the foundation of long-range ethics, because the direct experience of "the Other" is missing in long-range ethics. In the realm of food ethics the strategy of parallelism, i.e. acting *as if* or *by analogy with*, is not sufficient motivation for consumers to act according to their ethical values. A quick glance at the food market can convince anybody of this, for especially in the case of food long-range ethics are in a weak position. Gastronomy and the aesthetics of table manners are a conscious and cultivated distance from the organic origin of food. It is not unusual to think of the slaughtering of an animal as a killing and for that

[170] Ibid. p. 107.
[171] Jonas, Hans: *The Imperative of Responsibility*, p. 7.

reason many people wish to distance themselves from the natural origin of food. Eating can easily become too natural and too bodily.

The idea of founding long-range ethics on short-range ethics is right from a theoretical point of view but from a more practical point of view short-range ethics cannot just be applied in a similar way. If consumers are to put their ethics into practice, we must take the opposite direction. *Instead of extending the short-range ethics to the absent, the absent must be made present in glimpses*. This is the thesis on which the rest of this book is written. The intention is to let the production practice appear as a story, as the *production hi-story*.

The experience of the production history in glimpses exists in at least two versions. In the strong version consumers are given a direct experience of the production, although only in glimpses. They see parts of the production and the people involved at first hand, and therefore to some extent know the production history from personal experience. This is the case when consumers buying meat and vegetables from a farm know the farm from having visited it. Having seen how the farm is managed and how the animals are treated their experience will always be linked to the products from that specific farm.

In the weak version consumers know about the production from second-hand accounts. Johann Zedler (1706–1751) produced an *Encyclopaedia* (1732–1750) which states that "what you cannot experience yourself, in that you must follow the experience of others".[172] This is the communicated experience, an experience retold by someone who knows the production process from their own experience – in contrast to the supermarket setting where nobody knows anything about the production practices. Of course, if this communicated experience and knowledge is to have any value for consumers, the person conveying the production history must be trustworthy.

Both versions contain some degree of self-experience: in the weak version it is the experience of listening to a story told by someone who has knowledge of the production that we ourselves lack. In the strong version, the direct self-experience in itself is a guarantee of reliability, while the weak version requires the trust of consumers. As we shall see in the final chapter the personal trust, the meeting face-to-face with the production history, is important for the weak version.

In order to embrace consumers' taste for food ethics, both sensuality and reflection must be taken into account. Food ethics must be founded on both corporality and intellectuality. Sensuality and corporality ensure the proximity and presence, the face-to-face situation, so to speak, with the food. But that is not enough. Consumers must also reflect on the ethical aspects of the food production process. The idea of bringing this knowledge in narrative form serves to accentuate the ethical implications of the production in a general and easily

[172] Quoted from Koselleck, Reinhart: *Futures Past. On the Semantics of Historical Time*, p. 38.

understandable way. In the rest of the book the possibilities and limits of this narrative, semiotic, phenomenological and hermeneutic approach will be investigated in detail.

Of course, there are limits to how close consumers can get to the production history. The formulation "glimpsed" indicates first of all that it is not necessary to have experienced the whole food production history – which would also be practically impossible and would make the absent totally present and thus abandon the long-range ethics in favour of complete closeness. The aim here is a balanced position where the absent is not totally unknown, i.e. it is made more present without becoming pure presence but remain only as "closeness" or "nearness". Conversely it can be said that the present is extended to the absent without completely taking the place of the absent. Consumers have a certain distance to the production history, but retain a sensual and corporal relation.

When people have knowledge of the production history, the food in itself can, as we shall see in the next section, function as a clue to, or trace of, the production history. When we have a sensual experience of the food of which we have some knowledge, this knowledge can be made present by the sensing itself; the sensing of the food can be used to recall experiences or accounts of its production history. It is like a short flashback or a story that can be recalled just as fast as an idea. Food then plays a role in two stories that intervene in each other: the story of the food itself, its production history, intervenes in the personal life story of consumers. This kind of food ethics, developed by means of the trace of the production history, belongs to semiotics and narrative ethics.

2. FOOD AS A TRACE

Eating is not only about nourishing oneself to keep the metabolism going, to stay alive. Peter Farb and George Armelages state that food and eating more than any other essential behaviour bring people into broader contact with their total environment – not only their natural surroundings but also their social, economic and political relations with neighbours.[173] This may well have been the case once, but the distancing to, or cultivation of, food nowadays is almost entirely devoid of contact to the natural surroundings; nor does it contribute to the understanding of the social, economic or political consequences of our own food consumption. Today the art of cooking and the aesthetics of the table reflect a "will not to know", a more or less conscious avoidance of information about the origin and history of the food. The French structuralist Claude Lévi-Strauss (b. 1908) suggests that in our relation to food it is of less importance what is good to eat and of more importance what is good to think about.[174]

[173] Farb, Peter and George Armelages: *Consuming Passions. The Anthropology of Eating*, pp. 9–10.
[174] Lévi-Strauss, Claude: "The Culinary Triangle".

Food is less substance and more circumstance; it is more symbol than nourishment. Of course, it is absurd to deny the incontestable value of food as nourishment, as the substance of the metabolism. But if food were *only* nourishment it would not belong to the social and cultural sphere but exclusively to the natural sphere. When food is transferred from its natural origin to a social entity, when it is *prepared*, it belongs to the cultural sphere.[175]

Food that is good to think about is also good to eat; food that is unpleasant to think about is also unpleasant to eat. The way we think of food influences taste, sometimes even more than the physiological taste itself. Simmel draws our attention to the fact that it is polite to talk during the meal, but the issues under discussion should not disturb the pleasure of eating: conversation should at all costs avoid the mundane, biological origin of the food. The thought of something unpleasant disturbs the appetite; food is not only good to eat but also good to think of, writes Daniel Roche.[176] But it might just as well be the other way around, as the following personal anecdote depicts. When I think of the mouse that one cold winter's night was trapped in a mousetrap where it was fatally injured but with its dying twitch fell off the shelf and into a toaster where it died . . . And when I heard that the very same toaster on that very same morning was used to toast the bread that I was eating . . . my appetite slackened. Usually I am not touchy about food, but now and then this episode still disturbs me, for example the smell of toasted bread; initially it whets my appetite but then I am reminded of the dead mouse and my gorge rises. It is hard to tell how long these conflicting sensations will continue. But there can be no doubt that the smell of toasted bread will bring the crushed and dead mouse into my mind for years to come. Experiences in the past can be recalled by sense impressions; present sense impressions can put us on the track of former experiences. Knowledge of the production history cannot affect the physiological taste in itself, because it is not altered by any knowledge, but knowledge can have a powerful influence on the taste or gustatory sensation, that is, the way a taste is sensed or perceived. This is also where we find the very special link between taste and ethics: knowledge of the ethical dimension of the production history can influence the perceived taste.

This is a special kind of semiotics: *the food as trace of the production history*. The perception of food through the ability to recall experiences from the past offers an alternative to a food ethics based on labelling. Food as trace may of course lead the mind in many directions, but what counts here is to see how

[175] The preparation of food is often taken as the first sign of the civilization process. The "Peking Man" dating from some half a million years ago indicates that already at that time nature was not eaten raw but prepared. Farb, Peter and George Armelages: *Consuming Passions. The Anthropology of Eating*, p. 52.

[176] Roche, Daniel: *Histoire des choses banales. Naissance de la consommation XVIIe–XIXe siècle*, p. 240.

the trace can lead the thoughts towards the production history and in particular its ethical dimension.

"The special character of the trace", writes French philosopher Emmanuel Lévinas, "is that it means something without revealing what it means".[177] The trace does not show what it refers to but leaves its meaning hidden, and as such the trace remains a mystery. According to Lévinas the trace is distinguished from the sign in that it does not contain any intentionality; the trace does not intentionally *show* anything, it is signification without immediate reference to a meaning. Foodstuffs cannot function as signs but only as traces that let us know that there has been a production history in which there must be an ethical dimension. But as a trace the production history remains an enigma, because the trace does not reveal what kind of story it is, nor does it say anything about the people involved in the story. The causality of the food as a trace is limited to the fact that all food must have an origin in the form of a production history.

A completely different point of view is posed in the semiotics of the American intellectual Charles Sanders Peirce (1839–1914). Peirce was more interested in interpretation than intentionality and he defined the sign in the following way: "A sign is something that signifies something for somebody in one way or another".[178] The interpretation (in the above definition the interpreter is "somebody") plays a crucial role in Pierce's understanding of the sign. His concept of the sign has three components: (1) the primary sign, (2) the signified, that is, the object that the sign refers to, and (3) the interpreter. Following Peirce we can say that in the case of the trace interpretation is unclear: traces are latent signs because they *can* be interpreted, but they do not *have* to be interpreted and certainly not everyone can interpret them, for it takes considerable time and energy. Likewise it takes an effort from consumers to see and understand food as a trace of its production history, for previous knowledge of this is required.

Traces are typically some kind of residue or remainder that has been left from one activity or another. This relationship of the trace to an activity is just as characteristic as the lack of intentionality. The past is preserved in the traces that are left behind.[179]

The production history of the food belongs to the past, but the trace, i.e. the food, remains – at least until it has been eaten. As such the trace is the presence of something that is no longer present, something which has passed and cannot be changed, but which has not completely disappeared. The German philosopher Martin Heidegger (1889–1976) writes:

> A Dasein which no longer exists, however, is not past, in the ontologically strict sense; it is rather, 'having-been-there' [da-gewesen].[180]

[177] Lévinas, Emmanuel: "La trace de l'Autre".
[178] Cf. Hoffmeyer, Jesper: *Signs of Meaning in the Universe*.
[179] See Ricœur, Paul: *Time and Narrative*, Vol. III, p. 120.
[180] Heidegger, Martin: *Being and Time*, p. 432.

The past is present as *da-gewesen* and that is something completely different from saying that it has disappeared. Whatever has completely disappeared no longer has any influence on the present, but the things that have been (da-gewesen) and are not forgotten exert an influence on the present; the remembered past lives on in the present and into the future.

The trace is the entering of spatiality in temporality; it is the point where the world "bends" towards the past and earlier times.[181] The ethics of Lévinas are based on the concept of the Other and the meeting face to face. In the face of the Other, Lévinas sees the trace of the absent Other and he describes the relation between the signified and the meaning of the trace as one of *irrectitude*, as a kind of deflection towards the past. In his analysis of the trace in the face of the Other he writes that we only know the other person because we remember something about that person. We relate to the trace of the Other. Kemp interprets this trace as what we have left from what a particular person has said and done.[182] Lévinas is interested only in the traces left by human beings, not those left by other living creatures. Yet his analysis may nevertheless be also applicable in the case of food.

We should now ask the question how food can function as a trace of its own production history. Historians have their methods of reconstructing the past from documents and other items, but food cannot be compared to such documents. However, foods are "traces" in that there is a production *history* to them, not a vacuum. The position of historians and consumers is not the same of course. Even though consumers know that the food has a history, by and large they prefer not to know anything about it. As we saw in the genealogy of gastronomy the presence of the kitchen can be disturbing, because it is too close to the organic and natural origin of food. In consequence the kitchen and the dining room have been separated. Likewise today we separate production and consumption because the production history may be unpleasant, even ethically unbearable. Faced with the unbearable there are two options; to reject the cause or to strengthen oneself in order to be able to bear it. Translated to the food area this means either to stop eating (as vegetarians do not eat meat) or to become cynical and close one's eyes to what cannot bear the light of day. A third possibility is of course to make the production history more bearable, which is also the aim here: to unfold the ethics of consumers.

The events and the story that the food is a trace of belong to the past and are for that reason unchangeable, but like the hunter consumers can follow the trace of the food. In the ethics of food production this demands, in Paul Ricœur's words, that "we must be capable of that letting-go, that abnegation that makes

[181] See also Ricœur, Paul: *Time and Narrative*, Vol. III, pp. 120–122.
[182] Kemp, Peter: *L'irremplaçable. Une Éthique de la technologie,* p. 104.

care about oneself efface itself before the trace of the other".[183] To follow a trace takes hard work and requires that – as in any ethics that is not solely based on self-interest – we pay attention to and care for the world and the others that surround us.

In his book *Phenomenology of Perception* Merleau-Ponty often departs from the desk he is sitting at. In his "desk philosophy" he notes:

> This table bears traces of my past life, for I have carved my initials on it and spilt ink on it. But these traces in themselves do not refer to the past: they are present; and, insofar as I find in them signs of some 'previous' event, it is because I derive my sense of the past from elsewhere, because I carry this particular significance within me... but in no case will this perception [of the desk], which is present, be capable of pointing to a past event, unless I have some other viewpoint on my past enabling me to recognise it as memory. . . .[184]

The meaning of the trace stems from a common past. Lévinas talks about the trace as a deflection towards former events and the past. For Lévinas there is something about a face that leads our thoughts to what is left of what that person has said and done. One may object that in this case the face is no trace, in that it only reveals something for an acquaintance by bringing forth certain memories – the trace only reveals its meaning to an acquaintance who already has certain knowledge. The acquaintance is like the trained hunter that knows his prey from former encounters and is therefore able to find his bearings in time and space according to the tracks (traces) of the animal. In contrast, to the unskilled hunter the tracks suggest an animal has passed but not what kind. To the unacquainted person the face appears naked, though not as emptiness, because we know that the face hides a story. The same applies to foodstuffs, as also to other things: foodstuffs are traces because we know that there *is* a story, whether we are familiar with it or not. Like the hunter, ethically aware consumers must know something about the prey, they must be able to recognize and understand the traces, to reconstruct past events from these traces and to find their bearings. If this is not possible, the chance of losing their way is imminent and almost to be expected.

Just as the trace in Lévinas and Merleau-Ponty can recall the memory of what has been said or done in the past, so food as a trace has the capacity to evoke knowledge about the production history – if this knowledge is already there, i.e. if there is a direct or indirect shared past between consumers and the food. Food as a trace only reveals its meaning if there is already some knowledge or experience of the production history, just as the trace of the Other implies some knowledge and memory to be understood.

In recent years it has become more common to talk about the traceability of food. There are different ways whereby food can have traceability. Wine, for

[183] Ricœur, Paul: *Time and Narrative*, Vol. III, p. 124.
[184] Merleau-Ponty, Maurice: *Phenomenology of Perception*, p. 413.

example, is a kind of "food" that can be traced to the property where it has been produced – and in more costly cases to the exact field. Today traceability mainly refers to the ability of producers to trace the foodstuffs in the food chain. Traceability has become a central tool in the fight against food unsafety as it is used as a means to locate contaminants such as pathogens in the food chain. However, traceability does not necessarily have to be articulated in this way, it might just as well refer to something self-experienced, which is of course more sensual and corporal than pure intellectual knowledge.

Food as a trace refers to temporality. The trace always contains some kind of temporality. Ricœur argues that

> [T]o follow a trace is one way of 'reckoning with time'. How could the trace left in the space refer back to the passage of the sought-for object without our calculating the time that passed between them, that is, between the passage and the trace it left?[185]

Consumers always reckon with the past if they ask *when* the food has been produced. They also reckon with the future in asking for *how long* the food can be stored. To see food as a trace is therefore to place oneself in temporality. Foodstuffs refer to the past as a story of events, but they also refer to future events.

As was shown earlier, the rise of gastronomy was also a showdown with a certain kind of temporality, the tyranny of the seasons. Today we can say that technological development has completed this emancipation of the seasons. Temporality is a difficult matter within food production, since sophisticated methods of food conservation and the obscurity on the food market hide it. In the age of globalization it is always summer somewhere on earth and therefore there are always fresh strawberries on the market. Cheese with the special kind of odour coming from cows grazing on the meadow in the early spring can be made synthetically all year round. But the emancipation from the tyranny of the seasons has a reverse side too; food is no longer a trace of a local time. Consumers no longer experience any understanding or embeddedness through food in the local temporality. Merleau-Ponty writes:

> Time is, therefore, not a real process, not an actual succession that I am content to record. It arises from my relation to things. Within things themselves, the future and the past are in a kind of eternal state of pre-existence and survival; the water which will flow by tomorrow *is* at this moment at its source, the water which has just passed is now a little further downstream in the valley. What is past or future for me is present in the world.[186]

Likewise the animals and plants we are going to eat in the near future already exist in the here and now. From this point of view food ethics is more than just an "ethics of the past" (see Chapter One). As an engagement in, and care for,

[185] Ricœur, Paul: *Time and Narrative*, Vol. III, p. 124.
[186] Merleau-Ponty, Maurice: *Phenomenology of Perception*, p. 412.

those involved in the production history, food ethics can become an ethics for the food production of the present.

3. THE JUDGEMENT OF TASTE AND MORALITY

Before we follow the production history of food further, we must dwell for a moment on the trace to see how it reveals itself in sensuality and experience. As such it resembles other things that we experience through our senses, the main difference being that it has been somewhat overlooked as an important element for the individual's vital sensibility, "feeling for life" (*Lebensgefühl*), self-knowledge and cultural identity. At the same time sensuality in the trace contributes to the communication of a short-range ethics that stretches out towards long-range ethics, i.e. as an ethic for those areas of society and nature which despite their distance from me are not unaffected by my actions, and whom I therefore cannot ignore in my vision of the good life for and with others in the production of foods. Food as trace has a double nature; it is in close in the present as both sensuality and reference to the absent and to the past.

Sensuality in food is experienced primarily through taste, a concept that is used in two different ways: taste as the physiological sense located in the tongue and related to smell, and taste in the figurative sense of judgement, as analysed by Kant in his third critique, *The Critique of Judgement* (1790). The twofold usage of the word is found in many languages such as French *le goût*, Spanish *el gusto* and German *der Geschmack*. According to the German philosopher Hans-Georg Gadamer (1900–2002) it was first used in the mid-17th century by the Jesuit monk Balthasar Gracian (1601–1658), who saw in the sensual taste an ability to discriminate; taste *determines* whether something is edible or not, thereby situating taste between two poles. On the one hand taste is related to the sensual and instinctual drive, on the other hand discrimination relates the decision on edible or inedible to the freedom of the spirit.[187] Taste is at one and the same time a concept which, understood as a physiological sense, is disdained for its "naturalness" and its link with life in its most primitive, banal form, namely ingestion and defecation, yet is praised as the most sublime aesthetical judgement, as an expression of refinement and civilization. In this context a closer look at *The Critique of Judgement* can suggest how Kant contributes to our understanding of the perception of food partly through pure sensuality, partly through the role that the intellect and cognition play in relation to food.

In her book *Estetisk rasjonalitet. En analyse av konstitusjonsbegrepet i Kants Kritik der Urtheilskraft* (*Aesthetic Rationality. An Analysis of the Concept of "Constitution" in Kant's Critique of Judgement*) (1991) the Norwegian

[187] Gadamer, Hans-Georg: *Truth and Method*, p. 35.

philosopher Hjørdis Nerheim presents an original reading of Kant that contains a phenomenological and existential interpretation of the judgement. Kant's own very formalist, analytical language cannot be said to encourage such an interpretation; on the contrary, he is often regarded as a very speculative and intellectualized philosopher whose sensuality has been cultivated out of him.[188]

The Critique of Judgement is in two parts: (1) aesthetic judgement and (2) teleological judgement. The latter examines the purposiveness of nature as it appears to us and how it influences our judgement. In our case, however, interest focuses more on the first part and its treatment of the aesthetic judgement of taste, particularly on its reflections on pre-discursive reasoning and its relation to the intellect and rationality.[189]

Kant's argument is that there is a perception to be found in sensuality that is non-lingual, a perception that takes place *before* the intellect categorizes the perceptions of sensuality conceptually. *The Critique of Judgement* can therefore be read as an attempt to transcend the discursive compulsion of the intellect and the one-sided trust in rationality and reason of the Enlightenment.

Kant analyses the judgement of taste by first noting that as a subjective judgement it is aesthetic. Things appear before the subject in their sensuality and on the strength of the feeling of pleasure or displeasure, which they evince they give the subject the "feeling of life". Appearing before the subject in this pre-discursive fashion the world precludes cognition, for cognition belongs to the intellect, according to Kant. But it allows the subject to become conscious of his state of mind and his emotions,[190] and from this comes the feeling of life. In the case of food we can say that the sensuality of food is without cognition, but initiates a corporal presence. Spontaneously and without formulating the experience in words we know whether the picture is beautiful, the music is good or the pear tastes good. Taste, and more generally sensuality, are characterized by a general feeling of pleasure and existential well-being which contains no cognition and which cannot be captured in the discursive declaration of food contents. The pre-discursive nature of sensuality sets it in contrast to the discursive mediation of that declaration. By virtue of its intellectualizing the intellect is unable to establish an affiliation to the world. Nietzsche describes it thus:

[188] For example, Schmidt, Lars-Henrik: *Smagens analytik* (*Analytics of Taste*), p. 27: "This aestheticising of judgement is in reality an intellectualisation, which sets a possible *limitation* on judgement which modifies the imagination by distancing itself from the sensual and emotional judgement". Cf. also Nerheim, p. 89.

[189] The pre-conceptual content of sensuality appears in the following among others, Kant: *The Critique of Judgement*, p. 215: "But for this very reason the aesthetic universality attributed to a judgement must also be of a special kind, seeing that it does not join the predicate of beauty to the concept of the *Object* taken in its entire logical sphere . . .". Several times in the third critique Kant states that taste is not a perception in the form of concepts, e.g. also p. 287. Gadamer also sees taste as pre-conceptual: "Taste is therefore something like a sense. In its operation it has no knowledge of reasons". *Truth and Method*, p. 36.

[190] Kant: *The Critique of Judgement*, p. 204.

> The harsh daylight, rationality at all costs, life bright, cold, cautious, conscious, instinct-free, instinct-resistant: this itself was just an illness, a different illness - and definitely not a way back to 'virtue', 'health', happiness . . . . To have to fight against the instinct – this is the formula for decadence: so long as life is ascendant, happiness equals instinct.[191]

In the rational categorization of the world, where the intellect articulates the world, an scientific inter-subjectivity is created which is an intellectualized experience whose meaning horizon is limited. We sense that the longing of the reason for identity with itself is fulfilled through the aesthetic experience, says Nerheim.[192]

We saw in Chapter Two, on the natural history and the emergence of biology, that towards the end of the 18th century there was a shift in the importance of the signifier. The sign was no longer the thing itself, now it became a representation of the thing. We can compare this shift with the relation between the pre-conceptual character of the judgement of taste and the conceptual categorization of the intellect in Kant. The natural history view of nature is reminiscent of the pre-discursive sensuality with respect to corporality, intensity, presence and totality, but also of the limited capacity for categorization and control. The view of nature that emerges with biology builds on the modus operandi of the intellect; it imposes concepts and organizes, but is characterized by a fragmentary life-feeling and lacks the closeness of sensuality. At least officially, modern knowledge (or science) has rejected taste, the subjective judgement, replacing it with the instrumental view.

The argument can therefore be made that the Renaissance world picture is still valid in our time, that it continues to express a way of viewing the world and an existential being-in-the-world, an embeddedness in the world. But we will not find expressions for this within the language of empirical rationality, for the well-formed language of natural sciences is pervaded by representation. It is rather in poetry that we find a language that can express how the sign is the thing and the thing is the sign, for example in Peter Handke's poem *Lied vom Kindsein* (*Song on Childhood*) used in Wim Wender's film *Der Himmel über Berlin* (*The Sky Above Berlin*):[193]

> Als das Kind Kind war,
> ging es mit hängenden Armen,
> wollte der Bach sei ein Fluß,
> der Fluß sei ein Strom
> und diese Pfütze das Meer.
>
> Als das Kind Kind war,
> wuate es nicht, das es Kind war,
> alles war ihm beseelt,
> und alle Seelen waren eins.

[191] Nietzsche, Friedrich: *Twilight of the Idols, or, How to Philosophize with a Hammer*, p. 15.
[192] Hjørdis Nerheim: *Estetisk rasjonalitet*, pp. 71–79.
[193] Wenders, Wim und Peter Handke: *Der Himmel über Berlin. Ein Filmbuch von Wim Wenders und Peter Handke*. Suhrkamp, Frankfurt a. M., 1990.

Als das Kind Kind war,
hatte es von nichts eine Meinung,
hatte keine Gewohnheit,
saß oft im Schneidersitz,
lief aus dem Stand,
hatte einen Wirbel im Haar
und machte kein Gesicht beim fotografieren.

[When the child was a child'
he walked with his arms hanging down
and wished the brook was a river
and the river a tide
and the mudpool a sea.

When the child was a child,
he did not know he was a child,
everything was full of soul
and all souls were one.

When the child was a child,
he had an opinion about everything and nothing,
and he had no habits,
he often sat cross-legged,
ran away from the place,
had a cowlick in his hair
and made no faces when photographed.]

In his fantasy and play the child can well imagine that the brook is a river. But it is not as representation; the child *wills* that the brook *shall be*, or rather *is*, a river. The child has no distance to himself, nor to his surroundings. Things do not *mean* something, they *are*. Everything is animated and all souls are one. When the child is photographed he makes no attempt to pretend, for it would produce no particular "representation" or image of itself. It is as it is. Hannah Arendt's description of taste addresses this too. Taste and smell are unable to recollect objects because they are object bound, unlike vision, feeling and hearing which are able to represent sensations of the past). Taste and smell are subjective inner senses that cannot represent sensations of the past.[194] They are bound to the here and now and thus to being.

After Kant's description of the subjective side of the judgement of taste via perception Kant turns to the universal, "public" (*gemein*) element in the judgement of taste. From the outset he thus isolates the ambivalence of the judgement: its subjectivity yet its universality. The fact that the judgement of taste is universal means that it must be impartial, for the partial is not universal but private: "Everyone must allow that a judgement on the beautiful which is tinged with the slightest interest, is very partial and not a pure judgement of taste".[195] From which Kant argues that impartiality amounts to disinterest.

He then proceeds to analyse the meaning of "delight" (*das Wohlgefallen*) in the judgement of taste by examining three factors: the agreeable (*das Angenehme*),

[194] Arendt, Hannah: *Lectures on Kant's Political Philosophy*, p. 66.
[195] Kant: *The Critique of Judgement*, p. 205.

the good (*das Gute*) and the beautiful (*das Schöne*). He concludes that delight is not the same as agreeable; delight is linked to the agreeable and the good but it falls outside the judgement of taste, because one's attitude to these two is partial and affected by interest. Only in the judgement of taste as to what is beautiful is man's delight free and impartial.[196]

Kant thus employs the concept of beauty as the focus of his analysis of taste, because "The beautiful is that which, apart from a concept, pleases universally".[197] The choice is deliberate, for in the judging of beauty he finds both the subjective sensuality and a reference to the universal. The beautiful is not in the thing itself, it is a precondition, as Nerheim puts it, for "consciousness of a meaningful content in life".[198] Beauty lays claim to universality, in Kant's view:

> It would, on the contrary, be ridiculous if anyone who prided himself on his taste were to think of justifying himself by saying: This object (the building we see, the dress that person has on, the concert we hear, the poem submitted to our criticism) is beautiful *for me*. For if it merely pleases *him*, he must not call it beautiful.[199]

Kant's argument is that there is a *universality* that is *not* objective, because objectively nothing can be deemed "beautiful". The judging of anything as "beautiful" is a subjective category that does not rely on a postulate of everyone's approbation but which disposes everyone to approve the judgement of taste, thereby achieving universality.

The judgement of taste embraces both the subjective and the universal, says Kant. Kant justifies this through the strategy of drawing parallels, that is, by referring to two other areas: *in analogy with* or *as if* (*als ob*). In classical rhetoric a parallelization via a *simile* is expressed by *as, as if* or *just as* and acts as a meaning extension. In *The Critique of Judgement* Kant attempts gradually to develop and legitimize this view throughout the first part of the book by describing the dialectical element in the judgement of taste. We shall dwell on some of these parallels in what follows, especially where the analysis demonstrates phenomenological and existential features.

The judgement of taste, says Kant, has two qualities:

(1) "The judgement of taste determines its object in respect of delight (as a thing of beauty) with a claim to the agreement of *everyone*, just as if it were objective".[200] In the sentence, "This rose is beautiful", which Kant uses, lies not merely a subjective statement but also a universality. This is not the judgement of taste, as in "I like the taste of apples" which he also uses, for that is a private matter. It is the judgement of taste concerning beauty to

[196] Ibid. p. 210.
[197] Ibid. p. 219.
[198] Hjørdis Nerheim: *Estetisk rasjonalitet*, p. 27.
[199] Kant: *The Critique of Judgement*, p. 212.
[200] Ibid. p. 281.

which he is referring, an appeal to the statement-receiver that "It is not just me who thinks it is beautiful, everyone surely must do so". Physiological taste, in Kant's view, is unsuited to the judgement of beauty since it is bound to a personal interest.

(2) The other quality in the judgement of taste is that "proof is of no avail whatsoever for determining the judgement of taste, and in this connection matters stand just as they would, were that judgement simply *subjective*".[201] The subjectivity that is found in the judgement of taste is different from the sensuality that is linked to personal interest, as is the case with the appetite, for instance. Kant writes: "[W]e regard as coarse and low the habits of thought of those who have no feeling for beautiful nature (for this is the word we use for susceptibility to an interest in the contemplation of beautiful nature), and who devote themselves to the mere enjoyments of sense found in eating and drinking".[202] People with a healthy appetite think all that is edible tastes good:

> So far as the interest of inclination in the case of the agreeable goes, everyone says: Hunger is the best sauce. And people with a healthy appetite relish everything so long as it is something they can eat. Such delight, consequently, gives no indication that taste has any say in the choice. Only when men have got all they want can we tell who among the crowd has taste or not.[203]

Taste then is concerned not only with the beautiful, with the work of art and nature sublime – as can be seen from Kant's definition: "*Taste* is the faculty of estimating an object or a mode of representation by means of a delight or aversion *aside from any interest*".[204]

As we have seen above, the judgement of taste is not linked to private desire or interest, but on the contrary to disinterest. Kant describes the subjective sensation that produces a mental state (*Gemüthszustandes*) as having a universal capacity for being communicated (*allgemeine Mittheilungsfähigkeit*),[205] in other words something that each and every person can endorse and which therefore establishes a disinterested community.[206]

The anchoring of the judgement of taste in sensuality links it to "like" and "dislike", not as private feelings but as "disinterested desire", a kind of contemplative desire free of any interest. But how is one to conceive of a desire without interest, or as Kant puts it, a disinterested interest? To do so he makes use of another concept, namely "purposiveness" (German: *Zweckmäßigkeit*):

> From this there arises the concept of a *purposiveness* of nature, as a characteristic concept of the reflective judgement, rather than of reason, since the end is posited not in the object but rather always in the subject, and in fact in the latter's mere capacity for reflection. We call that "purposive", when its

[201] Kant: *The Critique of Judgement,* p. 284.
[202] Ibid. p. 303.
[203] Ibid. p. 210.
[204] Ibid. p. 211.
[205] Ibid. p. 217.
[206] Cf. Hjørdis Nerheim: *Estetisk rasjonalitet*, p. 8.

> existence seems to presuppose its antecedent representation; but natural laws which are constituted and interrelated as if the judgement had designed them to satisfy its own need resemble the case in which the possibility of a thing presupposes the representation of it as a ground. In this way the judgment, by means of its principle, thinks a purposiveness in nature in the specification of its forms through empirical laws.[207]

The idea of the purposiveness of nature is a principle that the reflecting judgement employs to give direction to the reflection, a way of systematizing and categorizing the chaotic impressions experience gives us. This is a principle that altogether enables experience to exist and is therefore a transcendental principle.[208] It is a characteristic of the subject that functions with the aid of an *as if*, and thus it does not represent any objectivity but an appearance (*Schein*) of reality.

According to Kant, expediency is linked to desire because it disposes the forces of consciousness to cognition. If we do not experience this consensual agreement that expediency gives to experience, we lose our desire because in the experience we feel things to be disconnected. This is the disinterested desire. Disinterest does not imply lack of interest but only the presence of the seeker in a state where no external aims exist.[209] Sensuality as found in the aesthetic judgement exceeds the controlling rationality of the intellect through the strength of its quest to grasp human experience and its totality. Nerheim writes:

> Life itself – the living life – is mediated *solely* as feeling. The judgement of taste involves a reflection so that one is led beyond the practice-orientated comprehension horizon of everyday reality in order thereby to respond to a silent meeting with life that precedes all immutable scientific understanding. In this sense the consciousness of taste – by virtue of its aesthetic judgement – is mute.[210]

Further definition of taste thus involves a desire to enter into a life from which we feel excluded from in our everyday humdrum. Nerheim understands *The Critique of Judgement* as an attempt to thematize the logic of closeness.

> When we accept the Other's judgement of taste, we also acknowledge his emotional consciousness of the world. But we thereby become capable even of seeing into the Other's emotions. We thus confirm the Other's deepest secret, which he experiences in the aesthetically pleasurable sensation, namely, that he is at one with his world. Through acknowledgement of his reflected-upon judgement on the beautiful the experience of an internalised I-You relation is fulfilled in such a way as to enable one to participate in the Other's life. The discourse on taste thus makes concrete the idea of man's most secret fellowship in relation to the world and each other.[211]

[207] Kant: *First Introduction to The Critique of Judgment*, p. 20 (Ak. Vol. XX, p. 216).

[208] Kant: *The Critique of Judgement*, p. 181.

[209] Hjørdis Nerheim: *Estetisk rasjonalitet,* p. 172.

[210] Ibid. p. 23. Gadamer also writes that "both taste and judgement evaluate the object in relation to a whole in order to see whether it fits in with everything else – that is, whether it is 'fitting'". *Truth and Method,* p. 38.

[211] Hjørdis Nerheim: *Estetisk rasjonalitet*, p. 142.

What then does this mean for our relation to food? There are two issues of major importance for the development of food ethics as a trace of the production history. Firstly, this somewhat elaborate detour via Kant's analysis of the judgement of taste suggests that the communication of the production history of a foodstuff based on sensuality and self-experience has deep existential perspectives which the merely discursive declaration of contents lacks. In the light of the above discussion on short-range ethics the importance of this perspective becomes clear. The sensual, exemplified in taste, links the individual to a real and existing world which is necessary for the ethical engagement. The sensual perspective is important for the positioning of food ethics in the here and now.

The second point may seem less surprising. It addresses the fact that judgements of taste, neither in the physiological or the broad sense of the concept, cannot be applied as a moral judgement on the food (but only as an aesthetic judgement). There are two lines of reasoning in this:

(1) Judgement depends, as Arendt writes, on the operation of the imagination, that is the ability to represent objects that are no longer present, and on reflection, that is on concepts. As shown above this is almost contrary to how taste works. Taste is based on feelings and is without concepts. There are, however, good indications that Kant himself thought that there were a relationship between the *interest in beauty* and moral:

> I do maintain that to take a direct interest in the beauty of nature (not merely to have the taste needed to judge it) is always a mark of a good soul; and that, if this interest is habitual, if it readily associates itself with the contemplation of nature, this indicates at least a mental attunement favorable to moral feeling.[212]

Kant's establishment of a relationship between the interest in beauty and morality is not as speculative as it seems at first sight. Aesthetics, as experienced in the beautiful and the harmonious, is often linked to the good. Beauty is good, which, however does not mean that the good is also that which is beautiful. Even so I do not think it reasonable to state that people with interest in gastronomy and tasty food also have a predisposition to a good moral attitude.

(2) According to Kant, the judgement of food does not belong in the same category as the judgement of the beautiful, for food is linked self-evidently to a purpose or intention and thus to an interest.[213] The judgement of food is not purely a matter of taste, for it is partial, according to Kant. This brings food into the sphere of ethics.

[212] Kant: *The Critique of Judgement*, p. 298. See Allison, Henry: *Kant's Theory of Taste*, for a thorough treatment of the link between taste and morality in Kant.

[213] Ibid, p. 230: "But the beauty of a man (including a man, woman, or child), the beauty of a horse, or of a building (such as a church, palace, arsenal, or summer-house), presupposes a concept of the end that defines what the thing has to be, and consequently a concept of its perfection; and it is therefore merely appendant beauty".

Indeed it was never my intention in this book to regard food merely as pure aesthetics and without a conceptual basis (even though that *can* be one way of regarding it); the idea of foodstuffs as a trace of their own production history reveals not only an ethical interest, but also that their production history cannot be recognized without the use of concepts and an intellectual approach. Thus, if we follow Kant, there cannot be any subjective universality in the judgement of taste in food, but the interpretation of a given production history will always contain an ethical element and will always relate itself to ideas of the good life – which the ethical consumers are of course interested in.

In Kant's view, the reason why we nevertheless to some extent can link food with "the beautiful" is not to do with its purely aesthetic sensuality but with the appendant beauty (*adhärirende Schönheit*) of the food itself. As we saw above, in this aesthetic judgement Kant speaks of a congeniality between sensuality and purposiveness which arouses the disinterested desire. The congeniality in ethical consumption is not solely aesthetic, it is not without concepts; nor in the final instance is it a universal judgement of taste. Judgement of food depends on at least two elements; the sensual and the intellectual. It is the agreement between on the one hand the good with and for others in the production history, and on the other hand good taste.

4. THE TRACE AS PRESENCE AND LOST TIME

In our daily lives we no longer have access to the kind of knowledge of the origin of foodstuffs or their production history that would make us reflect on their story if our attention was so directed. The situation in which the sense perception of food takes place is described above as the meeting between two horizons or two life stories, that of the food itself and that of the consumer. Our purely pre-reflexive perception has no access to a foodstuff's production history, yet knowledge of this can actually colour our taste experience. The phenomenological sense experience in question is not a pure and undiluted pre-personal perception but a reflected one – a perception reflected in the perceiver's own life story and experiences. What then is the path of this perception and reflection?

In *Phenomenology of Perception* Merleau-Ponty writes of the space and time perspectives of perception, taking the experience of a house as an example. Of spatial perception he says:

> [T]he house itself is not the house seen from nowhere, but the house seen from everywhere. The completed object is translucent, being shot through from all sides by an infinite number of present inspections which intersect in its depths leaving nothing hidden.[214]

[214] Merleau-Ponty, Maurice: *Phenomenology of Perception*, p. 69.

For Merleau-Ponty the experience of a house or any other object is not pure sense perception, for this will always see the house from a particular angle. Into perception is placed the added knowledge from experience, in this case that the house can be perceived from every imaginable angle. Sense perception is thus positioned in a spatial perspective that transcends the constraint of sense perception to a subject's position. And just as the house is seen from every angle, so is it seen from every time: the experience of time in the present is linked to the experience of time yesterday and the expectation of it tomorrow. The German philosopher Edmund Husserl (1859–1938) defines this with the concepts of *retention*, the reaching back of the immediate consciousness into time past, and *protention*, its reaching forward into the future. Retention and protention thus create a time unit through the synthesis of time horizons. But as Merleau-Ponty points out, the exactness of the synthesis decreases the further one moves away from the here and now. Things that happened long ago I may well have forgotten, just as my plans for the distant future are less certain.[215] In the synthesis of horizons the actual experience, the pure sense perception in the here and now, transcends its own time and space; things are perceived simultaneously from a thousand different sides (i.e. in their three-dimensional spatiality) and are comprehended in their time-context of past, present and future, and then things and events are placed in their singleness and wholeness as a synthesis in time and space. Sense perception is thus transcendent, since the limiting of pure sense perception to a specific angle and to the present is transgressed; and it is also a synthesis because in the final perception we collect many time and space perspectives in a single experience.

Sensuality resides in the body. For Merleau-Ponty corporality is important for an understanding of the path of perception. "I cannot", he writes, "understand the function of the living body except by enacting it myself, and except insofar as I am a body which rises towards the world".[216] This is the body, and to extend the definition to all living creatures we could follow Hans Jonas in calling it an *organism*, which reaches out to the world in order to meet its demands:

> The quality of sense – that the perceived spatial conditions and the actual presence or absence of a perception – is not the result of the situation outside the organism, but represents the way in which the organism meets the demands of the stimulation and relates itself to them. A stimulation is not perceived if it strikes a sense organ that is not 'tuned' to it.[217]

[215] Merleau-Ponty, Maurice: *Phenomenology of Perception*, p. 70.
[216] Ibid. p. 75.
[217] Jonas, Hans: *Das Princip Leben*, p. 89.

To return to food for a moment, we must acknowledge that it is difficult to follow the process by which food becomes foodstuffs. We believed that we could manage solely on the intellectualized declaration of contents. But this declaration stimulates neither the senses nor the intellect. The declaration has been unable to serve as a replacement for sensuality, corporality and the self-experience for which it was intended in an industrialized food production.

Let us look more closely at how consumers are confronted by a foodstuff. Through their glimpsed experience of the production history, the opportunity is created for the foodstuff in question to recall or *reopen the past*. Memory reawakens time. "All memory reopens lost time and challenges us to take command of the situation it evokes".[218] The reopening of the past means the calling forth of already "thought" knowledge that lies within the past situation and taking charge of it:

> When I chat with a friend whom I know well, each of his remarks and each of mine contains, in addition to the meaning it carries for everybody else, a host of references to the main dimensions of his character and mine, without needing to recall previous conversations with each other. These acquired worlds, which confer upon my existence its secondary meaning, are themselves carved out of a primary world which is the basis of the primary world.[219]

In contrast to the path of thought, which is relatively slowly processed into language, the amount of knowledge that memory can swiftly call upon is stupendous.

The knowledge or experience we possess can be recalled in the same way as an *idea* or as a collected *imaginative* thought. In a flash a knowledge and a mood are recalled without the former having to be retold. It is like saying "The Ugly Duckling"; immediately we know which story is being referred to, and we have the feeling for the story's train of events without needing paraphrases, let alone details. What sense perception evokes is not the complete articulated story, but an idea that is "recountable" (French: *racontable*).[220] But Merleau-Ponty also draws attention to the linguistic and conceptual understanding, which contains *acquired worlds*, as sedimentations of mental operations that enable us to master acquired complex concepts without recourse to repetition of their synthesis every time. Of course it is in this form of "conceptually acquired knowledge" that the declaration of contents is to be found, and as such it cannot be understood without prior knowledge.

So why then should an ethic not be based on this conceptual, acquired knowledge, if in systematized, standardized form it is capable of the same function and effect as the sensual trace? This is because, as we have already seen, the intellectualized mediation via conceptualization abstracts from the relation

218 Merleau-Ponty, Maurice: *Phenomenology of Perception*, p. 85.
219 Ibid. p. 130.
220 Cf. Ricœur, Paul: *Time and Narrative*, Vol. III, p. 260.

to the present, to self-experience and to being-in-the-world. Abstraction in its intellectualized mediation arises out of a lack of self-experience, which means that we must *imagine* the context we are acting in and we must act *as if* or *in analogy with* it, as we would have done if the production history had been present. But can we imagine this context? Even if the declaration is finely detailed, also with regard to information of an ethical character, it still remains an idea based on discursive information from the seller (and here the profit motive limits the trustworthiness). We can imagine a context without difficulty, but it is not certain that it matches the reality – and this is the weakness of the imagination. Conversely the strength of bodily experience is that what is now distant in time can be made mentally present in the here and now (though physically absent). The body's closeness to things, in the present or the past, prepares the consumer for an ethical engagement.

Since closeness takes on considerable significance for the consumers' food ethics we are led towards two practical solutions to ethical consumption: self-sufficiency, or some form of community production whereby the food is produced and consumed locally. For it is obvious that in the spatial, the absent can be made present by shortening the distance to it. The most radical and uncompromising solution would be to make the distance so short that total self-sufficiency is ensured, for then consumers would know absolutely everything about the production process; but of course this is impossible. There would no longer be a consumer problem, the distance ethic would have become a closeness ethic and the distance ethic would have no relevance.

The consumer perspective can better be preserved in a communitarian approach, where a smallish community trades within itself; some produce foodstuffs, while others consume them in a local context. But the local community is not an absolute necessity for closeness, we could just as easily imagine it being played out on the freer and more anonymous premises of the market. The market structure as such does not have to be impenetrable and globalized; this *can* be a consequence but it is not part of a natural law.

The problem of self-sufficiency and community production in relation to food ethics illustrates the danger in trying to reduce the ethical distance by shortening the physical distance and in emphasizing the ethic of closeness in order to make the ethics in question realistic and practical: In the attempt, the ethics of distance dissolve into too much closeness and indeed becomes a tyranny of closeness. For consumers there are problems in making the absent present in a spatial dimension in order to implement a food ethic. On the other hand time offers some opportunities to make the absent present without the physically absent becoming "purely closeness".

Part of the memory, says Merleau-Ponty, is linked to corporality:

> The part played by the body in memory is comprehensible only if memory is, not only the constituting consciousness of the past, but an effort to reopen time

> on the basis of the implications contained in the present, and if the body, as our permanent means of 'taking up attitudes' and thus constructing pseudo-presents, is the medium of our communication with time and space.[221]

A bodily memory, what does that mean? To reopen the past on the implications of the present is to let the sense impressions that the body has in the present lead the consciousness back to something in the past. I have already mentioned how a particular smell or taste or sound can for years after lead the mind back to a specific situation or person. In his novel *In Search of Lost Time* the French writer Marcel Proust (1871–1922) looks at the body's memory. Probably the most famous passage in the book is where a Madeleine cake quite unexpectedly evokes previous experiences and feelings:

> Many years had elapsed during which nothing of Combray, except what lay in the theatre and the drama of my going to bed there, had any existence for me, when one day in winter, on my return home, my mother, seeing that I was cold, offered me some tea, a thing I did not ordinarily take. I declined at first, and then, for no particular reason, changed my mind. She sent for one of those squat, plump little cakes called "petites madeleines", which look as though they had been moulded in the fluted valve of a scallop shell. And soon, mechanically, dispirited after a dreary day with the prospect of a depressing morrow, I raised to my lips a spoonful of the tea in which I had soaked a morsel of the cake. No sooner had the warm liquid mixed with the crumbs touched my palate than a shiver ran through me and I stopped, intent upon the extraordinary thing that was happening to me. An exquisite pleasure had invaded my senses, something isolated, detached, with no suggestion of origin . . . this new sensation having had the effect, which love has, of filling me with a precious essence; or rather this essence was not in me, it *was* me. I had ceased now to feel mediocre, contingent, mortal. Whence could it have come to me, this all-powerful joy? I sensed that it was connected with the taste of the tea and the cake, but that it infinitely transcended those savours, could not, indeed, be of the same nature. Where did it come from? What did it mean? How could I seize and apprehend it? . . . And suddenly the memory revealed itself. The taste was that of the little piece of madeleine which on Sunday mornings at Combray (because on those mornings I did not go out before mass), when I went to say good morning to her in her bedroom, my aunt Léonie used to give me, dipping it first in her own cup of tea or tisane. The sight of the little Madeleine had recalled nothing to my mind before I tasted it.[222]

The title, *In Search of Lost Time* (*À la recherche du temps perdu*) emphasizes precisely through "*in search of*" that this is not just a trace we are passively waiting for but an active quest for what has passed and is past. According to Proust this trace is to be found in the body. The human body contains the hours of the past; legs and arms are full of memories. The body's memory follows two traces: first, the reopening of the past through a sense perception, and second, the body schedule, understood as one's own understanding of the body's immediate spatial presence in the

[221] Merleau-Ponty, Maurice: *Phenomenology of Perception*, p. 211.
[222] Proust, Marcel: *In Search of Lost Time*, pp. 51–55.

world (the fact that I know, for instance, the position of my body without thinking about it),[223] and that certain body positions can recall certain past moments.

Of the time horizons (the past, the present and the future) Merleau-Ponty writes:

> The synthesis of horizons is essentially a temporal process, which means, not that it is subject to time, nor that it is passive in relation to time, nor that it has to prevail over time, but that it merges with the very moment whereby time passes.[224]

In the subject the fusion of horizons is experienced as a uniting of the past, present and future in "the moment", all three of which are present in the here and now. We must also note the intimate relation between the concepts of *presence* and *now* in French. The two meanings are covered by the French word *présent*. The present, *le présent*, denotes not just a time statement but also a closeness to things. Conversely, "closeness" describes a position in relation not just to an object but also to a contemporaneity, a here and now. Another French word for *now* is *main-tenant*, which can be understood as *what I hold in my hands*. Again we see how the present in French is linked to spatiality, and vice versa. In addition to covering the closeness of time and space, *présent* in its figurative sense also means presence of mind, openness and receptivity in both body and soul. In French, therefore, it is not too much to say that consciousness in the present, *le présent*, is ubiquitous in relation to things. Existence occupies this ambiguous position, as *le présent* suggests: it is a basic condition "the present" is inseparably linked to both time and space.[225]

As we shall see with Gadamer, experiences do not stand alone, isolated from other events, but are fused with other experiences in both the present and the future. To claim that the present is not influenced by the future and the past – that the past does not continue and the future is not there – belongs to the linear time-view of modern natural science. Existence as found in the *Confessions* of Augustine (354–430),[226] is related to three distinguishable times – the past, the present and the future – but the present is not isolated; rather it is coloured by events of the past that we carry within us as memory. The past is the background upon which the events of the present play themselves out and into which they inscribe themselves. The future is close at hand in the present as idea and expectation of what is to come. Dreams and plans for the future affect our experience in the present, serving to direct the experience to the here and now. Experiences are thus far from being isolated events or episodes, but exist within

[223] Cf. the description of the body figure in Merleau-Ponty, Maurice: *Phenomenology of Perception*, pp. 98–101.

[224] Merleau-Ponty, Maurice: *Phenomenology of Perception*, p. 330.

[225] Ibid. p. 397.

[226] Cf. *Confessions*, Vol. XI. Analysed by Ricœur in *Time and Narrative*, Vol. 1.

a context of past and future, which despite their separation from them are nevertheless present in the present, so to speak. Memories and expectations gather the past and the future in the present. Augustine illustrates time as such through the artistic form, which par excellence relates itself to time, namely music. In music we can "hear time" in a way:

> I am about to repeat a psalm that I know. Before I begin, my attention encompasses the whole, but once I have begun, as much of it as becomes past while I speak is still stretched out in my memory. The span of my action is divided between my memory, which contains what I have repeated, and my expectation, which contains what I am about to repeat. Yet my attention is continually present with me, and through it what was future is carried over so that it becomes past. The more this is done and repeated, the more the memory is enlarged – and expectation is shortened – until the whole expectation is exhausted. Then the whole action is ended and passed into memory. And what takes place in the entire psalm takes place also in each individual part of it and in each individual syllable. This also holds in the even longer action of which that psalm is only a portion. The same holds in the whole life of man, of which all the actions of men are parts. The same holds in the whole age of the sons of men, of which all the lives of men are parts.[227]

Augustine presents the specific character of time through music, and this leads us to a general problem which may be said to be contained in the mysteriousness of time,[228] that is, the problem of measuring and describing time. "What is time?" he asks: "When no one asks me about it, I know what it is; if anyone puts the question to me and I wish to answer it, I no longer know it".[229] His inability to describe time tempts us to turn to comparisons and parallels. If time cannot be explained through itself, it may perhaps be understood or at the least be illuminated by something else. Thus we find in Ricœur's *Time and Narrative* a linking together of time with narrative, which examines how time becomes human through narrative.[230] Like music both history and the narrative are played out. The actions and events of history and the narrative insist on a time dimension to make sense. One can barely imagine events and actions that are not played out in time.

Let us now see how via time the lack of closeness and engagement in the ethics of distance can be overcome. It is here that the glimpsed self-experience of the production history plays a part. The situation we must imagine is thus that the consumer can now and then experience the production history in glimpses. Via memory the absent can be made present. The absent – what was previously experienced as present – can be called forth and become a presence of the

[227] Augustine: *Confessions*, Book 11: 28–38.

[228] In brief, according to Ricœur in *Time and Narrative*, Vol. 1, Augustine describes time as mysterious, because it is both "there" and yet is "not a thing". This raises the question of how to measure it and in general how to "think" time.

[229] Augustine: *Confessions*, Book 11: 14–17.

[230] Ricœur, Paul: *Time and Narrative*, Vol. 1, p. 52.

absence. For the subject the absent is present, not in a spatial sense, but in time, as a reopening of time.

In the actual moment where the production history is self-experienced, the absent is totally absorbed by the present – the absent *becomes* totally present. Perhaps the consumer has travelled many miles to experience the production history, and later may even wish to repeat the trip. But in our day-to-day life the production history is spatially absent and we are geographically separated from it. Through memory, however, we can recall that specific production history to the present moment as "the closeness of something absent". The foodstuff in question is the object that serves as a trace that pervades the senses and the body and reopens a memory distanced in time and space. The memory "leaps" in time and space and can evoke the doubleness of experience: the present time with all its sense perceptions of its immediate surroundings and that time's occupation of space existing alongside the closeness of the thing remembered. We are in two places at the same time, so to speak, even though they are literally separate in time and space. Physically of course it is quite impossible to be in two places simultaneously, so the distance is in a sense retained. But while we are "here", memory can carry us "elsewhere", thus sustaining the "presence" of the "absence". Merleau-Ponty describes this as the contradiction of consciousness or a paradox that memory is ubiquitous yet engaged in the present.[231]

The foodstuff as trace appeals as a stimulation to the senses, to corporality. It is a more powerful mediator than the discursive, "scientific" declaration of contents. Merleau-Ponty writes:

> I become involved in things with my body, they co-exist with me as an incarnate subject *(sujet incarné)*, and this life among things has nothing in common with the elaboration of scientifically conceived objects.[232]

According to Merleau-Ponty, in perceiving we do not think of the object, nor of ourselves as thinking, but we are *with* the object, uniting with it through our body, which knows more about the world than ourselves.[233] Perception is centrifugal, because the subject is directing his attention outwards to his surroundings. Thus to understand a thing is always to construct and constitute, and to undertake an actual synthesis of the thing.[234] In German it is *Sinn-gebung*, giving meaning to things (*Sinn* meaning simultaneously sense, consciousness, meaning and reason). Perception is neither a passive nor objective reception of impressions from without, but a subjective and synthesis within time of previous experiences and the present. Sense perception in this sense is a contrasting opposite to the declaration of contents, for it tends not to objectivity but is individually subjective. The sense

[231] Merleau-Ponty, Maurice: *Phenomenology of Perception*, p. 331.
[232] Ibid. p. 216.
[233] Ibid. pp. 238–239.
[234] Ibid. p. 429.

perception of the present is coloured by each person's previous experiences, that is, the life story that has already been run, and every sense perception inscribes itself into this flow of former sense perceptions and experiences.

The "glimpsed" production history that was mentioned before makes it possible to extend the ethic of closeness to the ethic of distance. In the outward orientation and openness of the body and the senses lies an engagement in the world. Foodstuffs as traces give a sensual and bodily relation that strengthens the experience of unity in the world.

As traces of foodstuffs are representations; they re-present the production history that they themselves are a result of. As we saw with the emergence of biology, the change of the sign to becoming a representation also meant an intellectualization. When we are to relate to a production history we do so not with sensuality but with the intellect and reason. Foodstuffs as they confront consumers acquainted with the production history contain this doubleness of sensuality and reflection. The re-presentation (the trace) appeals to the intellect, while the foodstuff itself appeals to sensuality. Foodstuffs as traces are at one and the same time intellect and sensuality. For the ethics of food it is doubtless important that the production history is something that can be experienced, but to understand the ethical element in the production history, it must also be a narrative.

5. HISTORY OF EFFECT

Hermeneutics concerns itself traditionally with the interpretation and exposition of texts. It is a humanist art of interpretation. The exposition of the Bible, exegesis, belongs to classical hermeneutics, whose methodology was formulated during the 16th century in connection with reformation theology.[235] With the German romantic Friedrich Schleiermacher (1768–1834), who emphasizes individuality and the individual's possibility for expressing his feelings, hermeneutics is extended to a more universal *doctrine of comprehensibility* within the humanities. Gadamer also writes from within this hermeneutic tradition; in his book *Truth and Method* from 1960 he describes Schleiermacher's emphasis on living feeling rather than cold rationalism as a protest against the incursions of the industrial society. And we can also see the romantic hermeneutic if not exclusively as a counter-movement at least as an alternative to the faith in enlightenment, rationality and positivism in the headlong rush to natural science.

Gadamer applies the hermeneutic to the field of art and tradition, where he underlines it as being an *understanding* of a text, a picture, or in general, a work. When it becomes a hermeneutic discipline even to understand art,

[235] See Gadamer, Hans-Georg: *Truth and Method*, p. 171 ff., for the historical development of hermeneutics.

aesthetics simultaneously becomes an important part of the hermeneutic. But as with the Schleiermacher this is not enough. He seeks to understand the psychological background of the artist's expression by returning to its origin, that is, to the originator or the artist. This suggests, as Gadamer writes, a historical dimension, which in Schleiermacher is still not clearly expressed, but which is nevertheless the tacit prerequisite upon which Schleiermacher builds his hermeneutic and which is to bring the lost back to the light. In continuation of this Gadamer interprets the traditional theological and philological hermeneutic, which disciplines to

> a rediscovery of something that was not absolutely unknown, but whose meaning had become alien and inaccessible. Classical literature, though constantly present as material for humanistic education, had been completely absorbed within the Christian world. Similarly, the Bible was the church's sacred book and as such was constantly read, but the understanding of it was determined, and – as the reformers insisted – obscured by the dogmatic tradition of the church.[236]

For Gadamer the context into which the work is produced is important for its understanding. He describes the hermeneutic consciousness as a tension between familiarity and foreignness. The attempt to understand presupposes that something is unknown and the unknown or the foreign, which also has its own history, must be met with an openness; it must be welcomed in before it can be understood. On the other hand, says Gadamer, every experience happens in a subject who has already had a number of other experiences; the perception of the unknown does not happen in an empty, unprepared subject. Gadamer expresses this using the word *prejudice* (German: *Vor-urteil*). Normally, the word has a negative ring, meaning a too-hasty and/or false judgement, but Gadamer seeks to rehabilitate the concept by pointing out that there are both legitimate and illegitimate prejudices. The former are those pronounced before the final judgement, a pre-judgement (as opposed to a false judgement). For Gadamer these prejudices represent a continuity, a kind of preservation of what has happened which cannot be escaped and which is present in spite of all the changes that the narrative has undergone and will continue to undergo in the future.[237] As such the prejudice builds on the memory and on the traces, which the narrative has set. We carry our prejudices within us and they colour our experiences of the still unknown; that is the horizon we ourselves see and which may be great or small. Understanding plays out between openness in the moment and the prejudices we carry with us from the past:

[236] See Gadamer, Hans-Georg: *Truth and Method*, p. 171 ff., for the historical development of hermeneutics, p. 174.

[237] Ibid. p. 270 and 282.

> *Understanding is to be thought of less as a subjective act than as participating in an event of tradition,* a process of transmission in which past and present are constantly mediated.[238]

The hermeneutic has its place "between the past and the present", and it is not so much the task of hermeneutics to develop a method that achieves an understanding as it is to describe the conditions under which that understanding takes place.[239]

With the concept *history of effect* (German: *Wirkungsgeschichte*) Gadamer links understanding to the events of the past. A consciousness of the effects of the past is a consciousness that the past, the historical, plays a role in forming the understanding. Despite the distance in time that historical episodes occurred, we still carry them within us and they have an *effect* on our understanding of the events of the present. In our reading, the history of effect appears as reflection on the tension between the text itself and the present time in which it is read. The history of effect is linked to the understanding of the concept of *situation*. It is characteristic, says Gadamer, that one does not stand *facing* the situation, one stands *in* it. One is in a situation where one faces something, but where one also carries within one's own historical consciousness.

The above can be illustrated by an example of a hermeneutic exposition of food. Bearing in mind all the stories with which we surround food it is reasonable to see them as part of our relation to food and foodstuffs as such. This is what the English psychologist John L. Smith has done in his book *The Psychology of Food*

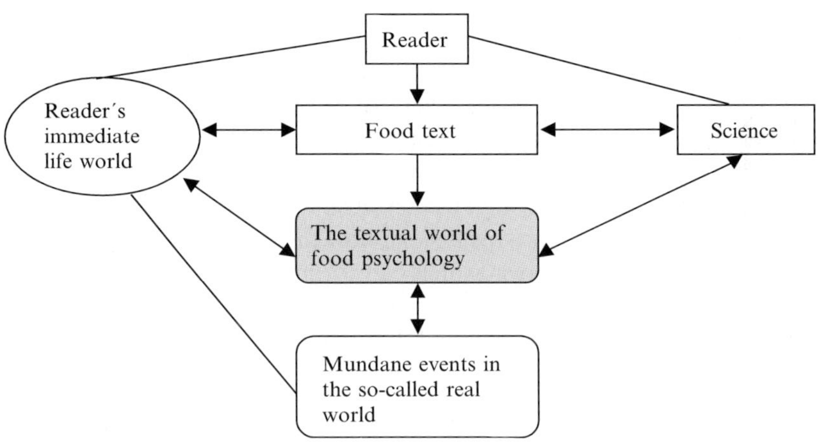

Figure 1. The discursive world of foodstuffs from a psychological point of view.[240]

[238] Ibid. p. 290.
[239] Ibid. p. 295.
[240] See Smith L. John: *The Psychology of Food and Eating*, p. 203. The figure is here simplified on the right-hand side, where Smith differentiates between various psychological disciplines.

and Eating. Drawing on Ricœur as his starting-point he has developed his own model for the discursive understanding of food and eating (see Figure 1).

Smith's model is an analysis of man's understanding of food from a psychological viewpoint. Academic psychology (to the right in Figure 1) takes the form partly of scientific texts on, and partly of experiments (empirical psychology) with food. The point is that texts and experience from experiments serve to create the reader's (here the psychologist's) own discursive understanding of food and foodstuffs. Smith's model is interesting because of its methodical attempt to regard through a hermeneutic approach, namely to build up the understanding from single parts which exist in both the past and the present and form them into a unity.

There is no doubt that the written word affects our relationship to food, but it is as if Smith is caught up in the discursive approach, unable to move into the phenomenological and existentialist relation to food. Smith admittedly operates with "events in the real world" and "the reader's life world" in his model and assigns them great significance, though without detailing what this life world actually mean. Thus he believes that professional knowledge "at the end of the day. . . will be evaluated in the last resort against the "reader's immediate life world".[241] But Smith's analysis of the life world tends towards the objectivity of natural science: those who are eating are considered in isolation, without relation to the world around them, as though they are objects for study in the closed-off laboratory of psychology. His analysis demonstrates a lack of reflection on the relation of eating to the social world, to space (nature), time, sensuality and corporality and to existence.

The reason for drawing attention to the hermeneutic here is of course because foodstuffs have a production history that can be put into words as a narrative to supplement the formalized information in the declaration of contents. Knowledge of this production history presents the opportunity both to reduce the amount of information and through the narrative to engage consumers in the ethical aspects of the production. This history, now turned into narrative, can create an understanding in the receiver which can form the foundation for an ethical viewpoint. The declaration of contents can also be seen as a narrative, though too brief and defective to serve as a basis for an ethical judgement. Let us imagine that we have set out to tell the production history of a foodstuff. If we are not careful, we will soon be merely presenting a list of facts which at best contains something about the place and method of cultivation, animal welfare, processing, working conditions, environment certification and contents of substances. It will be the story of the production of a foodstuff from its origin and doubtless overwhelmed with technical details, with little sense of the time needed for the

[241] Smith, L. John: *The Psychology of Food and Eating*, p. 208.

maturation of the product and few if any personalities presented, with no plot to drive the story on and with little narrative cohesion. Such a "story" will leave little impression, excluding as it does the possibility of sympathetic insight and identification. On the contrary, it is a story that bores us with its uninspiring technical details; we never become part of this story because we cannot identify with any people in any roles. The farmer and the baker could be any old farmer and baker and could just as well be a butcher or a candlestick-maker.

It is also a story without a development in time, for it can be repeated ad infinitum day after day, year after year. Nor does it occupy any space; it could just as well be here as elsewhere since it is not bound to a specific locality. It is history without a plot. People, time, locality and action play little role in the history of foodstuffs. However technically refined the production is, it is a terribly banal history of cultivation, processing and distribution that means nothing to us.

In contrast one could tell a production history with real people in real time in a real place, perhaps even close by. If this narrative is to have an impact, the narrator must know the production history. In the supermarket, where personal assistance is limited to payment at the till, there is no one as a rule who knows the production history of any of the articles. But when personally experienced, or related by the butcher or the fisherman, the narrative takes on a practical relation to the life world, as we shall see in Chapter Five.

This leads us back to sensuality as a point of entry for knowledge of the production history, which through this narrative has become a story. With regard to the story of food in relation to physical taste it is a fact that the story does not need the taste. The oatmeal does not need to become porridge for its story to be told. This may sound banal, but we shall return to the point in a moment for further comment. The fact is that the story of the product exists independently of the consumers, and they in turn can relate ethically to the production history without ever having eaten or intending to taste the product. Conversely the taste does not need the story for food to taste good; nor can the production history be deduced from the taste of the food. Indeed the physical taste may be the same for two widely differing production histories.

It makes no sense therefore to speak of a link between taste and the production history, if we are moving solely in one direction, as in one-way communication. The coupling of sensuality to narrative only makes sense if we regard both simultaneously: the sense perception and the narrative must exist simultaneously to be able to influence each other. This is dialectical in the sense that sense perception and narrative exist and act in a reciprocal relation. Knowledge of the history colours the taste, so to speak, creating a mood that affects the experience. And vice versa. More than anything else physical taste, and in a broad context sensuality, call up the stories and information that the memory has stored. Food thus becomes an opportunity to relate to nature (food is cultivated nature) and social space. But because the foodstuffs themselves as traces do not reveal the production history but present themselves as inscrutable, it is easier just to give up on the trace. And of course there are many who have an interest in foodstuffs being cheap – never mind the "story" – because they cannot afford for them to be otherwise. This of course includes many consumers.

But if we persist, like the hunter following the trace, we will find two guides to the events of the past, to the history. For whoever wishes to act ethically, it is not just a question of reconstructing the production history. The aim is to put the consumer in the historian's shoes to reconstruct the history and then interpret the narrative. It is the reconstruction of the history as action, which can be interpreted ethically.

In speaking about the past historiography makes use of sources of all kinds, from rubbish and tools found at archaeological sites to glaciers and plant remains to documents and archives. The sources are isolated and then brought together in "collections". The French historian Michel de Certeau (1925–1986) calls this process the "constitution of data", the material being the traces of former actions.[242] The gathering of source material is not solely an objective process but something which historians influence through their choice or rejection of sources: according to Certeau, the collection creates or recreates history. In creating these stories of the past historians are also creating a new reality; in reconstructing the past we reshape the present.

Foodstuffs differentiate themselves from most other traces of history by revealing nothing about themselves when regarded in isolation. Not until it is placed in a "collection", i.e. together with other sources, can the foodstuff become a trace. It thus requires several sources – just as historiography must build on several sources in order to build up a coherent and believable story. It is through being put together with other sources that the trace reveals its significance, and the collection into which it is put can comprise both documents and experiences. We do not need to have seen and experienced how those eggs or that cabbage came into being; the important point for foodstuffs to function as traces of their production history is that we either have glimpsed it or have a reliable source who has. My use of the word "trace" here is different from the way historians use language sources such as documents; for these direct them more or less explicitly to something else and reveal it, whereas in my usage a trace is a footprint for the memory, making the absent present. This usage is closer to that of archaeology (though without totally coinciding), where the individual finds do not as a rule immediately reveal their significance, but where their importance only comes to the fore on the background of a string of traces, a collection indeed, which together form a complete picture. The trace is not a revelation, but a mental evocation of specific knowledge. The foodstuff

[242] Certeau, Michel de: *L'écriture de l'histoire*, p. 84.

as trace cannot reveal the production history (for either we know it or it is unknowable to us), but it can orientate our thoughts in a particular direction.

The glimpsed experience of the production history, whether direct or mediated, thus becomes important for an ethical stance, for here we meet the people and the nature involved, making it a representation or a trace of something that is past, namely its own production history. The strength of the experience in an ethical context is that it indicates immediacy and closeness:

> Everything that is experienced is experienced by oneself, and part of its meaning is that it belongs to the unity of this self and thus contains an unmistakable and irreplaceable relation to the whole of this one life.[243]

Following Gadamer we can say that the meeting with food in the act of eating is a meeting of two narratives and two horizons: on the one hand is the production history of the food itself, on the other are the ideas and expectations from our own experiences and life story with which, via the history of effect, we interpret the production history. It is in this twofold experience that understanding arises. In a sense Gadamer's "pre-judgement" corresponds to what I have called the personal life story, and pre-judgement as such has already arisen from the life already lived, the experience that comes with experiences. If we apply the hermeneutic concept of understanding and try to "understand" the foodstuffs – which is literally nonsense but which figuratively denotes the meaning that the production process has for nature, man and society – then the foodstuffs must be brought into this field of tension where the two horizons meet and the personal life story crosses the trace of the production history.

With the foodstuff as trace the production history becomes present for a moment, remembered in a brief glimpse because it leads the memory to its story and offers the opportunity to exercise an ethic of distance. The closeness of the moment through the presence of the foodstuff makes us inhabit the world and gives us the "feeling for life". The absent is not entirely absent; it is physically absent but mentally present since it has been experienced previously, whether directly or mediated, and is thus present as a memory.

The previously glimpsed experiences of the production history merge with the consumers' own life story, embedding themselves in the memory of that story. The experience enters the context of other people and their surroundings. Consumption of foodstuffs acquires a direct link to other people's lives and to nature. The question of who produced this apple does not disappear into a vacuum but can be answered concretely, with a reference to specific people and geographical areas. This of course does not guarantee care and responsibility, but it is a first condition for being at all able to exercise it.

[243] Gadamer, Hans-Georg: *Wahrheit und Methode*, p. 67.

As mentioned, the absent is not entirely dissolved in the present. It is precisely a question of balance, where the absent can be present via the memory. The retention of distance, that is, the distance to the absent, is important because it contributes new dimensions to the assessment and interpretation of the production history. The moment the experience moves into the past and writes itself into a narrative of something that has happened, one also acquires a distance to what has been experienced. We can be present in the narrative, but we can also move out of it and regard it from a distance in order better to see the broad outlines. So in the production history as narrative we are simultaneously close by and yet at a distance.

6. PRODUCTION HISTORY AND MIMESIS

If we see the production history as a narrative, then to understand the ethics in the narrative we must ask how it should be told and interpreted. By way of answer I turn to Ricœur's *Time and Narrative* (1983–1985), where he develops a concept of mimesis in three parts. As in Aristotle mimesis means a staging of the action of a drama, a creative imitation of man's actions.[244] Ricœur speaks of three different forms of mimesis, which he calls mimesis I, mimesis II and mimesis III:[245]

> Mimesis I is the use that the narrative makes of the field of action in everyday life, from where it gains material for its account. This is the narrative's pré-figuration, its pre-understanding of the practical life. We only find meaning in the narrative because we are already familiar with the meanings that are given in the everyday world we live in, where we realise what it means that someone acts, sets goals, uses means, enjoys success or suffers defeat, etc.[246]

The reason that we understand the narrative is that through our practical experiences we have already built up an insight into what the concept means. According to the Danish anthropologist Marianne Horsdal (b. 1946), who writes from an educational perspective, the beginnings of narrative competence are acquired already in childhood.[247] Her thesis is that the genesis of narrative is bodily and belongs together with movement in space. But actions are not solely confined to space and concrete situations, there is a time dimension to them. Time and the narrative, as Ricœur points out, are closely linked. And in

[244] Ricœur, Paul: *Time and Narrative*, Vol. I, p. 31.

[245] The present account of Ricœur's mimesis-concept is based on Peter Kemp's interpretation in *Tid og fortælling. Introduktion til Paul Ricœur*, pp. 34–58.

[246] Kemp, Peter: *Tid og fortælling. Introduktion til Paul Ricœur*, p. 35.

[247] Horsdal's use of the concept *narrative competence* in *Livets fortællinger – en bog om livshistorier og identitet* is closely related to mimesis I, while in Peter Kemp's analysis Ricœur uses it to formulate narrative sentences and place these within a text that is other and more than the sum of the sentences; to establish a coherence, i.e. configuration; and finally to explain a historical event through dramatising it (Kemp, Peter: *Tid og fortælling. Introduktion til Paul Ricœur*, pp. 41–42).

the book *Women, Fire and Dangerous Things* (1987) following Aristotle's observations on the narrative George Lakoff (b. 1941) states that bodily experiences are often played out as a beginning (starting-point), a number of stages (the middle or the turning-point) and finally an ending.[248] Even quite simple actions such as drinking a glass of water follow this model. The starting-point is thirst, the middle is filling the glass and the ending is drinking it and quenching that thirst. Lakoff's point is that this bodily experience (which may very well be unconscious) of actions unfolding as a beginning, middle and ending is used in our understanding of the narrative, since both follow the same pattern. Experience teaches us that actions cannot be isolated from, but must be related to, one another, since a present action builds on past events and has future consequences. It is also a bodily experience when something fails or succeeds. And with this we are approaching the relation of ethics to the narrative.

> Mimesis II is the creative activity that composes the 'action' of the narrative in accordance with what Aristotle calls 'necessity and reasonableness'. On the basis of the order which the narrator sets from the start, what happens must be a necessary consequence of such a kind as to make us believe that it is 'appropriate' or 'reasonable'. Establishing this narrative order means, as we have seen in Aristotle, that there is a beginning, a middle (or a turning-point in the plot) and an ending. This is its configuration . . . . It is this activity that unfolds in the writing of both history and literature.[249]

First, Horsdal describes the ordering of the narrative elements as a selection and a hierarchy, where questions must be asked as to what must be included or discounted, what is important or unimportant for the story. Second, there must be a definite system in the story's sequence; the events, characters and circumstances, etc. must be presented in a meaningful order. Kemp writes that the narrative must be more than just a rattling off of events; these must be organized into an understandable totality. It must be possible to follow the development of the narrative. Thirdly, Ricœur states that the action must be set in a time-frame and through unfolding "poetically" must be able to cross the border between chronological and existential time:[250]

> Mimesis III completes the mimetic process, in that we do not experience the configuration of the text unless we appropriate it. Something happens to us when we 'follow the story'. We become 'new people' through the narrative's refiguration of our lives . . . . The narrative opens up a world that we can live in. It reconstructs our field of action.[251]

Mimesis III is our empathy with the narrative, where we set ourselves in place of the characters in order to experience the world from their standpoint. This opens up new horizons and new ways of understanding the world. In the story the

[248] Lakoff, George: *Women, Fire and Dangerous Things*, pp. 275–278.
[249] Kemp, Peter: *Tid og fortælling. Introduktion til Paul Ricœur*, pp. 35 and 46.
[250] Ricœur, Paul: *Time and Narrative*, Vol. I, p. 128.
[251] Kemp, Peter: *Tid og fortælling. Introduktion til Paul Ricœur*, p. 36.

readers' world intersects the story's world, two horizons meet and fuse. The readers' own horizon is not just their own experiences, but also the texts they have previously read. A special tension field is to be found between the fictional story and historiography. The latter borrows the productive imagination from fiction to reconstruct the past, and conversely fiction borrows its referential dynamic from historiography's reference to the traces, which as we have seen show that something has passed by; in this way the narrative is told as though it really had happened.[252] Ricœur calls this the cross-reference between fiction and historiography.

Mimesis can thus be described with three processes: pre-figuring, configuring and finally refiguring. From the narrator's view pre-figuration unfolds in the past (beginning), configuration in the present (middle) and refiguration in the future (ending). In what follows I will draw on Ricœur's mimesis-concept to examine how this threefold analysis of the narrative can contribute to an understanding of the production history. Prior to this, however, it is essential to know the situation in which the production history finds itself, for whether the self-experience is direct or mediated it is not a traditional writer-reader or speaker-audience situation we are dealing with. Three questions must be asked: Who speaks the narrative, who receives the narrative and which narrative is it?

To start with the simplest point, it is obvious that the receivers of the production history narrative are the consumers. My claim is that their ethical influence can be exerted through a glimpsed experience of the production history.

Who then is the narrator? Clearly the consumers' own experience of the production will inform the production history as narrative. Indeed, as eye-witnesses they themselves become in a sense narrators of that history.[253] When they experience the production history and make it part of their own life story, they become both narrators and audience. But there are also other narrators. The glimpsed self-experience of the production brings consumers into contact with producers, who through their knowledge and daily management can make a significant contribution to the production history as narrative, being simultaneously both narrators and actors. Visiting their workplace one often meets them as living narrators face to face.

A third narrator is the seller, whose knowledge of the production history is expected but not always found – for example in supermarkets – and whose integrity is essential for consumers if they are to believe the production story.

Finally, there is the question of which narrative to tell. The production history is an account of how the foodstuffs were manufactured, of animal welfare, slaughtering, environmental concerns, working conditions, processing, transport and much more. It is thus a narrative which, dealing as it does with real

[252] Ibid. p. 55.
[253] This is the sense in which MacIntyre uses the concept of history in *After Virtue*, p. 215: "What I have called a history is an enacted dramatic narrative in which the characters are also the authors".

events in the past, resembles historiography rather than fiction. But "narrative" is not quite the right word in this context, for it is not necessarily a narrative articulated through language. It is a retelling that does not always require words, so long as the trace can jog the memory and bring it to consciousness – before all linguistic formulations are applied and time-consuming reflections set in. It is like a song that we know by heart and to which we stretch out in its entirety, calling up its specific tone and atmosphere and focussing on specific passages that are characteristic. Other passages may not be open to instant recall without their context within the totality (even indeed when one knows the piece by heart). It is often seen that musicians cannot just start in the middle of a musical piece, say at bar 37, they have to come to it in its context, as part of an entirety. This is particularly true of musicians outside the western context, who do not subscribe to the analytical approach that divides a music totality into smaller sequences. They may have a "total picture" of the piece, but more as a feeling than as the individual details. The total impression shows itself in the fact that they cannot move in and out of all the details, for these hang together, the one paving the way for the next. In a somewhat similar fashion the foodstuffs whose history one knows beforehand (perhaps even by heart, as in music) can be "summoned up", due to a pre-narrated feeling for the story in its entirety. We link the egg to the memory of it, perhaps as inner images of the hens in the henhouse and the farmer tending them or from the time we were told fragments of the production history. But these pictures stand for a totality; if we want the details we shall have to play the piece and tell the story.

As a rule the experience of foodstuffs will be no more than this general impression, which steers clear of details but which is nonetheless of significance. Experiences and memories gather in the body and the consciousness as hi-*stories* and narratives, also even if they are not necessarily formulated in language. Experiences can also be remembered as pictures, actions, sounds, feelings, atmospheres etc., but they are narrative inasmuch as they contain the possibility of becoming language for that purpose. Once a narrative has been told and experienced it no longer requires words to become a conscious memory; a single trace can set off the narrative in its entirety and as an idea.

The production history can thus become part of the consumers' own life stories. Under all circumstances it can link up or intersect with them, entering their totality of events, situations, persons, etc. Food is an ever-present entity, but a "silent document", witnessing to the spatially absent life of production. The production history, as it appears to consumers who have glimpsed it, is thus a special story, since it links up with their own life stories.

Returning to the mimesis-concept we must now ask how the pre-figuration, mimesis I, relates to the production history. Primarily it is the general understanding of what "narratives" are, and as such it will be known to all consumers. On the social and human level, it is reasonable to assume that the ability to identify with the production history is sufficiently developed because we are familiar with the significations which the narrative recounts. But in relation to nature and the production organisms we must be careful not to assume that consumers have a pre-understanding from practical life. In a country such as Denmark, where 85% of the people have grown up in towns, nature and agriculture play only a minor part in everyday life, and there are many who have only very limited bodily experience of them. In Chapter One we saw how food ethics are to be understood as the consumers' relation to the producers (including the farmers), to the production organisms and to nature. One would therefore expect that the relation to them in the production history of foodstuffs would not find such resonance with consumers as the social aspect.

This seems not to be the case, however, according to the conclusion of the Nordic Council of Ministers' report from 2001 with the title *The Consumers' Sense of Ethics*. Ethics are diffuse for consumers, the report concludes; ecology is what comes closest to ethical understanding, especially care of the environment. Social care as expressed through consumption is hindered by being restricted to what is close at hand and is difficult to extend on a wider scale. The diffusion is due to the fact that, as we have seen, the complexity of the market makes production impenetrable to consumers, in contrast to when they have experienced the production history in some way. On the ethically diffuse market it is easier to formulate ethical concerns for nature, as more or less standardized ecology, than it is to implement practical care for people. There are several reasons for this. In addition to the problem with the long-range ethics in acting as if in complex contexts, a diffuse market, where the primary concern is to earn money, means that it is difficult to create relationships of trust and to show care for, or solidarity with, those whom consumers do not trust. In such cases solidarity remains with oneself. The starting-point for the experienced production history is therefore not the diffuse market but the close bodily experience, and social care is more likely to take precedence over care for the production organisms and nature. All things being equal, forcing children to work is worse than forcing hens to lay eggs.

On the market the declaration of contents offers information, but insight into the production history is more than just information, it is also a narrative from which consumers attempt to draw a meaning. Here too we can expect that meaning to be easier to see in the social area than in nature and the production organisms, since the common experience in relation to the latter is reduced at the price of the social care. The expectation of meaning arises from the preceding context, which for consumers today is primarily social.

So much for the pre-figuration. How then is the production history configured by and for consumers? How is the order created which helps to present the production history as a narrative? As regards their life stories, their autobiographies or psychoanalysis, consumers are both senders and receivers. Whoever tells their life story or parts of it can hardly avoid interpreting it at the same time; indeed, the interpretation most often accompanies the choice of what to include or reject. In psychoanalysis the patients tell their stories and the analysts interpret and lead the analysis in the interpretation of the narrative.

The first important element in the experience of the production is the staging, the presentation of the narrative's figures, i.e. the producers, but also of the consumers themselves. We need to know about the scenes where the narrative unfolds, its social environment and natural surroundings.

The production history also has an organized sequence of events, though not on the level of a novel or a drama, more like a life story, and as such, it is real, not fictive. My fundamental assumption is that the glimpsed experience is important for the ethical action, and thus the production history, like the life story, is a moment of self-experience for consumers, becoming part of their own life stories, a bi-narrative, however minor, that blends with their own personal life narratives.

Hannah Arendt defines events as occurrences that interrupt routine processes and procedures.[254] In this sense the glimpsed experiences of the production history are precisely an event of special character; although not decisive moments they nevertheless differ from the everyday routine and become easier to remember.

These life stories are less structured in their composition than literary works, according to Horsdal; the selection of material is less considered and coincidences can easily arise. They are not necessarily chronological, and associations may well lead the narrative into bi-narratives that perhaps have no real beginning, middle or ending.[255] This is also the case with the glimpsed production history, where consumers do not know the entire story but only fragments of it. In such cases the face-to-face meeting with the producers is of vital importance, furnishing them with a feeling for the totality of the story, partly because the producers can narrate the areas of production that cannot be seen, partly because by their very presence they can create an atmosphere of trust – or not, as the case may be!

The production history will have various narrators, some of it being self-experienced by the consumers, some narrated by the producers or the seller, often without a single beginning, middle or ending but with several stories fusing with each other. To avoid this sounding like the rattling off of a list the production history must therefore remain a narrative. The glimpsed experience of the production history gives consumers the opportunity to tell the story themselves – to create a beginning, a middle and an ending – and thus themselves undertake the selection of what is important in order to be able to organize

[254] Arendt, Hannah: *On Violence*, p. 12.
[255] Horsdal, Marianne: *Livets fortællinger – en bog om livshistorier og identitet*, p. 114.

history in an understandable totality. The organizing of the production history as a narrative, and not merely as a list, is an attempt to underline the ethical significance at the cost of, for example, technical details.

Modern food production is a complex business, and the question is whether consumers are willing to understand the complicated processes that are involved in food production, such as the use of enzymes and the actual content of gene technology. Complexity is the worst enemy of food ethics, because it renders ethical action impossible and the consequences of any action thus become incalculable. It is therefore also the case that the more foodstuffs are manufactured, the harder it is to take up an ethical position on them, since the demands on the individual's knowledge increase accordingly. The story becomes more inscrutable as more and more players and independent stories are gathered in the narrative. The more complex the process is, the harder it becomes to select the events and descriptions that are important from an ethical viewpoint. The market is today so complex that it appears ethically diffuse, as a narrative without beginning, middle and ending. For food ethics as seen by consumers it is absolutely central that the production history can take the form of a narrative. This does not mean, however, that the production history cannot bear a certain degree of complexity, or that we should return to the country kitchen of 150 years ago, which was so simple that it must almost be designated inadequate. The complexity can be reduced by waiving unnecessary information, as indeed is the principle with much labelling, though it is important for consumers that nothing is hidden. It creates consumer confidence to know who selected the information and on what basis, even though customers may not avail themselves of this opportunity, for it forces producers to provide comprehensive information.

Consumers inscribe themselves in a specific time-relation to the production story. The glimpsed experience means that the production exists not only in a chronological time but also in existential, experienced, time. Though not themselves present at the actual production consumers can experience it as if they had been, because they know that the foodstuff in question has been produced in the same way as it was glimpsed.

We appropriate narratives by reliving them. Narratives become appropriated through being relived. In the process of refiguration we understand ourselves anew: "The narrative work is an invitation to see our practice as if . . .".[256] Horsdal writes:

> [N]arratives of the life that is lived [one's life narrative] are a repetition, a reproduction of the path of events which seeks to give it meaning reflexively by making a selection, a hierarchy and a sequencing – and thus creating a

[256] Ricœur, Paul: *Time and Narrative*, Vol. I, p. 155.

> plot – from a particular narrative position that demarcates the course of events. When we try to imagine what will happen in the future, we test various configurations in imagined scenarios on the basis of the experiences we have from our narrative position.[257]

Horsdal is here arguing that the refiguration is not merely a reorientation of our lives based on the past or a fictive narrative, but a return to our past to see whether the story could be told in another way. One could object that refiguration always does this in that it relates to what is said – in other words it takes up a stance in relation to what are perceived as good or bad actions. This is true of course, but in self-experienced narratives, where the narrator and the interpreter coincide, where one is both author and reader, the opportunity to consider whether the narrative could or should have happened in another way is brought much closer. The narrative of the production history is changing continuously because it is seen in the light of the development of our own life stories, which offers new understandings and perspectives of the past. As narrator we must consider to what degree our new versions are repetitions of what actually happened or are reinterpretations. For the ethical perspectives the crucial point is that via the retelling of the production history consumers ask themselves what could or should be done differently and better. The refiguration fuses with the configuration to see if the past could have been different – for it cannot be changed – and whether through possible changes in production the narrative of the future can be better configured.

[257] Horsdal, Marianne: *Livets fortællinger – en bog om livshistorier og identitet*, p. 39.

CHAPTER FIVE

FOOD ETHICS AS THE ETHICS OF THE TRACE

We can present foodstuffs as the trace of their production history – or not, as the case may be. The foodstuffs are in a sense foreign to consumers, and it requires both will-power and insight to see them as traces. Nietzsche has a somewhat unusual passage that reminds us that we do not just see things as they are, but that we learn to see them from a particular perspective:

> *One must learn to love.* – This happens to us in music: first one must *learn to hear* a figure and melody at all, to detect and distinguish it, to isolate and delimit it as a life in itself; then one needs effort and good will to stand it despite its strangeness; patience with its appearance and expression, and kindheartedness about its oddity. Finally comes a moment when we are *used* to it; when we expect it; when we sense that we'd miss it if it were missing; and now it continues relentlessly to compel and enchant us until we have become its humble and enraptured lovers, who no longer want anything better from the world than it and it again. But this happens to us not only in music: it is in just this way that we have *learned to love* everything we now love. We are always rewarded in the end for our good will, our patience, our fair-mindedness and gentleness with what is strange, as it gradually casts off its veil and presents itself as a new and indescribable beauty. That is *its thanks* for our hospitality. Even he who loves himself will have learned it this way – there is no other way. Love, too, must be learned.[258]

It is the same with the production history: it must be learned. First of course we need to become aware that foodstuffs actually have a production history, though that in itself is not enough. For we need also to open ourselves to that history and make room for it, identifying with its characters, its action and its consequences, etc. But since this history is not self-explanatory, it demands a modicum of goodwill from consumers to acquaint themselves with it. Finally comes the moment when they are conversant with it and empathize with the players and their motives and can perhaps even see the consequences of the actions played out in the production. They may even come to know the history so well that they are surprised when it does not show as expected; indeed they may even come to "miss" it, since they have come to appreciate it.

[258] Nietzsche, Friedrich: *The Gay Science*, Book 4: St Januarius: 334.

On the other hand, we can do the opposite; we can ignore the foodstuffs as trace, actually prefer to remain ignorant and in this sense act "unethically" in not wishing to take a stance. This is unethical in the sense that we are choosing not to know and refusing to commit ourselves to even making a choice. Of course there are situations where we cannot justify making an ethical decision on a particular production history, since it would be out of place or not add anything new or good to the situation. If we insist that food ethics must play a role in absolutely everything we eat, we are liable to irritate both ourselves and others because we have different views and approaches to the subject.

Food ethics therefore has a particular part to play when we find ourselves in the role of shoppers shopping. For it is when we are buying foodstuffs in the stores that food ethics is worthy of consideration. From the consumers' point of view food ethics is thus linked to a specific situation and identity, not as the diner in good company but as the purchaser of the food.

Foodstuffs will always make up a trace of their own production history despite our ignorance of it, for like most things they always leave behind a trace of their origin. Of course our observation of an object, say a foodstuff, does not per se lead us to an instant recall of its production history; for that to happen two special triggers are required. The first is our relation to that object in the past, a common history or experience with it. The second is our relation to what we regard as historical sources or materials: documents of a past age, archaeological objects (traces of human activities) and remains of living organisms or a geological development (the history of nature, the biological evolution). Merely by looking at such objects we link them immediately to our historical understanding.

Conversely, we are much less inclined to see the historical dimension in the everyday utility objects that surround us, for they are there primarily to be *used*. Marx writes:

> As the taste of the porridge does not tell you who grew the oats, no more does this [production] process tell you of itself what are the social conditions under which it is taking place, whether under the slave-owner's brutal lash, or the anxious eye of the capitalist, whether Cincinnatus carries it [the process] on in tilling his modest farm or a savage in killing wild animals with stones.[259]

The everyday articles in question are first and foremost tools and instruments used purposively in the here and now, in the present; consumers are not expected to research their history before making use of them. And yet the future will be interested in their *history*, even while we are considering them purely for the purpose of forming the present and the future. Just as their character as everyday articles overshadows their history, so does our use of foodstuffs in the present and our storing of them for the future overshadow their

[259] Marx, Karl: *Capital*, Pt III, The Production of Absolute Surplus-value, ch. VII, The Labour Process and the Process of Producing Surplus-value, sec. 1.

production history. We are concerned first with satisfying our present and future hunger, and only thereafter with our aesthetic taste. At this level history is barely a concern.

Hunger in particular makes us forget that foodstuffs *have* a history. But where there is adequate food, there is a chance that the existential fear of hunger and the concern for one's immediate needs can give way to a solicitude that stretches out beyond the individual self to the other person.

If foodstuffs are to manifest themselves as more than insignificant traces of a production history, a trigger link must be made to their history. This contact must be to a display in time and space; it takes time and it fills space. In a sense the trace is like a symbol; it does not immediately reveal its hidden meaning, but can be deciphered through prior knowledge. If the receiver has no knowledge of the symbol's meaning, it is literally meaningless for that person. On the other hand, with the receiver who understands the symbol's meaning a link is set up to a specific presentation, which it re-presents. The Eucharist is a ritual and symbol of Christ's body and blood, but only for those who are aware of this. It is like a photograph of a person that has a special meaning if the viewer has a relation to the subject.

In this chapter I wish to focus on how this relation can be set up to the production history of a foodstuff, creating an identity for it among consumers and mediating a relationship to nature. Before doing so, however, there is an insistent question to be considered: how can we *trust* the trace? In other words, is the production history when it is mediated to the consumers a trustworthy one? Or is it merely a false trace leading us in the wrong direction?

1. THE RELIABILITY OF THE PRODUCTION HISTORY

Foodstuffs function in a market context – they are traded. In so far as "modernity" in a sociological sense can be characterized by the rapid increase in European trade from the end of the 17th century, foodstuffs are "modern". As a trading object they are important for relations between people and are crucial to the social economy. Exchange is fundamental to society, for it creates a relation between the involved parties and thus a specific social order. In the liberalist tradition, the commercial relationship between a seller and a buyer is in principle a relation in which both parties act in their own interests[260] and both benefit from the transaction, so both are satisfied. They share a common interest in reaching an agreement, but in the fixing of the price they have conflicting interests. Most often it is the seller who knows most about how the product was made and what the constituent manufacturing processes have cost. The seller

[260] Cf. Misztal, Barbara A.: *Trust in Modern Societies*, p. 33.

can increase the profit either by raising the price (but perhaps thus limiting the sale) or by reducing production costs. On the other hand, as we have seen, it is *cheapness* that is important to the consumers, who in their ignorance are afraid of being cheated, while simultaneously cheapness is also a determining parameter for producers who wish to sell. When consumers know that they are trading in a free market and under free competition, and that cheapness is a vital factor in being able to sell at all, there is good reason for consumers to assume that production costs have been minimized. That is of course a sound principle, which consumers must also support – but only within certain limits. This then is at the heart of the concerns and of the ethical dilemma that faces consumers: on the one hand, production must be rational and cost-effective; on the other hand, producers are expected to pay ethical regard to their product. The two principles pull consumers in opposite directions; finding the right balance between too little and too much of them constitutes food ethics in practice.

The reason that a foodstuff can be used as a trace of a social relation is because we know that the production history involves people in work. The precondition for the consumers' ethical choice must be care for, or solidarity with, the producers and the production workforce. If one is indifferent to them, there is no need for ethical reflection on them. Ricœur's definition of ethics in *Oneself as Another* is, as we saw in the introduction, "the good life with and for others in just institutions".[261] "With and for others" indicates the close social context that the good life occupies, whereas "institutions" refers to people in their organized lives in society. A fundamental understanding of solidarity must be that one is not alone but exists in various contexts with other people.

But do we as consumers *need* to show solidarity with foodstuff production? Can we not just leave it to those who are better judges of it? Is it not enough for consumers to show solidarity with themselves?

There are several reasons why the answers to the last two questions are "no". First, there is simple self-interest. We can surely calculate that if we want foodstuffs of a particular quality, we must also give agriculture and other food manufactures the conditions that enable them to produce according to quality and ethical criteria and not merely for the sake of cheapness. There is thus a reciprocal relationship of dependency between consumers and producers. Complexity is the first obstacle that food ethics must overcome, and the purpose of the glimpsed production history is precisely to reduce this complexity so that the mutual dependency becomes visible.

So it makes sense for consumers to show solidarity with producers; the question then is how can they assume that responsibility with practical consequences? It is not their job to produce or to organize the production, for as consumers they are not involved in it but are separated from it. Their only responsibility is to be

[261] Ricœur, Paul: *Oneself as Another*, ch. 7.

able to say: "It's me who consumes the foodstuffs produced in this or that way". The mutual dependency between consumer and producer means that consumers cannot be without any responsibility – though admittedly they cannot be charged with the whole of it. Consumers are involved with the product and its history precisely by *being* consumers. We saw in Chapter One that the common meal can be an unrivalled way of instituting a community or friendship. Food is shared, consideration is shown in a social context, all the consumers of the meal are involved. Here people do not eat one another but share the food and show one another consideration. The food in question directs the attention to the fact that one is eating not alone but in a social context.

Now of course producers and consumers do not dine together, and the consumers' ethics cannot be linked to the production via this route. Nevertheless, in a sense both are *present* at the meal, the consumers by virtue of the producers' production, and the producers through the price the consumers have paid; so even though they do not work in the same branch they are sort of table companions. The English *com-panion* and the French *co-pain* derive from Latin and mean, "someone one eats bread with".[262] It is unusual to regard consumers and producers as "companions", since on the market one man's meat is often another man's poison, but in the case of food ethics it is important to realize that consumers and producers do in a sense eat bread together. To refuse to share is to exclude ethics.

Thus, consumers form part of the production context and have good reason to show their solidarity with it – the problem is that they lack an overview and are therefore often referred to others' mediation of the production history. The glimpsed self-experience of this history provides a sensual dimension which can increase credibility, but there are many factors in the production history that consumers will be unable to evaluate and where they must be referred to the statements of others. As a rule, consumers cannot assess what constitutes a rational, efficient, professional production. However, the ethical reflection on production practices will often appeal to a universal understanding of "the good life", e.g. working conditions and animal welfare. The reliance on the information and testimony of others rests on a balancing of two elements in the evaluation of what constitutes a "reasonable" production when weighing them up: efficiency and ethical concern. Because they lack the producers' first-hand knowledge and expertise, *referral to others is a basic condition for consumers.*

This in turn raises the question of the reliability of the information sources. Can the consumers trust that the information given is correct? The answer to this question rests to a large extent on who is the sender of the information, i.e. the question is in reality *who* can the consumers trust? The reliability of information is estimated by an evaluation of the sender. Trust is a key concept within

[262] Farb, Peter and George Armelages: *Consuming Passions. The Anthropology of Eating*, p. 4. and Goften, Leslie: "Bread to Biotechnology; Cultural Aspects of Food Ethics."

food ethics, since consumers tend to rely on information given by sources they trust and to reject information from sources they distrust. Information is not evaluated equally by consumers but depends on the origin of the sources and on the consumers' knowledge of them. Knowledge and trust are thus closely linked for consumers. However, trust is linked not only to their assessment of the reliability of the sources and the information they provide but also to their own ignorance. Without uncertainty or ignorance the concept of trust would be superfluous. In his book *The Philosophy of Money* (1900) Simmel writes that people who know everything do not need trust. On the other hand, a certain degree of knowledge or familiarity is required to show confidence. Trust extrapolates from the available evidence and is therefore a blending of knowledge and ignorance.

One of the few, but very penetrating, analyses of the concept of trust has been carried out by the German sociologist Niklas Luhmann (1927–1998) in his essay *Trust. A Mechanism for the Reduction of Social Complexity* (1968). As the title suggests, his central thesis is that "trust is required for the reduction of a future characterized by more or less indeterminate complexity".[263] In Luhmann's later works he argued from another perspective that a reduction of the complexity is vital, since it is impossible to communicate on everything; within politics, ecology, law and the media, for instance, the complexity of "reality" must be reduced in order to make it comprehensible. When an issue appears so complex that it takes an extraordinary effort to understand it or to gather sufficient information about it, there is always the option of leaving it to others and trusting that they can do it better than oneself. This explains why the importance of trust increases in modernity; since science and technology cannot apparently bring matters under control (which could have rendered trust superfluous), replacing trust with mastery is not possible. On the contrary, trust is increasingly in demand.[264]

In Luhmann's view, one fundamental condition for displaying trust is the possibility that it may be abused; indeed there must be a *considerable interest in doing so*![265] Unscrupulous producers have an interest in deceiving consumers, for instance by lowering production costs illegally – of course at the risk of being revealed and sanctioned. Trust only appears when the involved partners are aware of the option for a breach of confidence, but they reject this opportunity. It should be clear from the previous chapters that the food market tends to be dominated by distrust rather than trust. The dominant distrust relieves the producers and consumers of social ties and moral obligations and gives them the freedom to act only in their own interests. In situations domi-

[263] Luhmann, Niklas: *Trust. A Mechanism for the Reduction of Social Complexity*, p. 15.
[264] Ibid, pp. 15–16.
[265] Ibid. p. 42.

nated by distrust one can expect that self-interest, which inevitably disappoints the other, will be prevalent. In that case complexity returns to the actors and their actions, since they can no longer be reduced or simplified by means of trust. At once the need for more information returns. Those who distrust are faced with the task of gathering and reworking the mine of information inherent in the complexity of modernity. Breaches of confidence also often lead to sanctions, countersanctions, etc. Strategies of combat, of mobilization of reserves or of renunciation make possible a conduct that is even *based* on distrust.[266] The sanction of specific products by consumers is precisely a strategy of renunciation and thereby distrust. Producers and retailers use "cheapness" as a combat strategy, the enemy being the competing producers and retailers, the losers in the end being all of them: consumers, retailers, producers, animals and the environment.

Let us return to Luhmann's analysis of trust. His approach is matter-of-fact, bordering at times on the inhuman; he does not consider trust from an empathetic point of view, or as a question of solidarity, as does Barbara Misztal in her book *Trust in Modern Societies*. Trust, says Luhmann, is a conscious renunciation of information and control. It contradicts the style of trust to ask for detailed information: "In the last resort, no decisive grounds can be offered for trusting".[267] Trust, and distrust, strengthen realities in a way that is difficult to trace back to specific causes. For this reason, Luhmann states that trust rests on illusion, as there is not enough information available to give grounds for action or assurance of success; yet we choose to believe (or trust) that things are going to turn out right anyway. To trust means deliberately to ignore the lack of information. Trust rests on the truster being already au fait with certain general features, being already informed, however inadequately and unreliably. It is the willingness to take a risk that could have been avoided if we had all the necessary information. Trust is a jump into the unknown, an operation of the will, since we can choose to trust somebody or not. By introducing trust, certain development scenarios can be excluded from consideration, and certain dangers, which cannot be removed but which should not disrupt action, are neutralized.[268]

When we trust, we admit to our own vulnerability as the instrument whereby a trust relationship may be created. Showing trust exposes our dependence on others and opens us up to abuse. On the other hand, whoever wants to win trust must be able to accept the expectations of others and act according to these expectations. We must be able to build the expectations of others into our own self-presentation.[269]

[266] Ibid. p. 71.
[267] Ibid. p. 26.
[268] Ibid. p. 25.
[269] Ibid. p. 62.

It is not possible to show trust in people who merely act according to rules or orders. To show confidence in somebody demands that the trusted person must be able to deviate from rules and orders. This calls for flexibility from the trusted person, who must remain free to decide yet at the same time able to incorporate into their actions the expectations of others.

Returning to the production history and the question of whom to trust on the *mediation* of this history – and thereby assessing what information can be considered true – we can acquire some clues from Luhmann's hermeneutic analysis. As consumers it is impossible to know everything about the production history; indeed consumers are mostly left with *no* means of evaluating the truth of the information given. They must therefore be selective and discriminating about the sources of information they use. This in turn demands first of all a certain foreknowledge for the evaluation of the trustworthiness of the sources. Such prior knowledge can be in the form of an acquaintance with persons involved in the production history, a personal trust, or an acquaintance with organizations or systems, a system trust. Personal trust is based on personal relationships, whereas system trust tends to be based on factual and technical expertise, e.g. on internal rules or control mechanisms of the system. Where personal acquaintance, the face-to-face relation, is impossible, system trust can function as a replacement of personal trust. The English sociologist Anthony Giddens (b. 1938) considers the separation of time and space – the "lifting out" of social relations from local contexts (also called disembedding) – to be an essential characteristic of modernity, which is also the condition for the need and emergence of system trust.[270] As we are no any longer bound by the local place and time, we are forced to rely on abstract systems.

However, there may be another side to the concept of system trust. Nietzsche pointed to a lack of honesty and integrity as a basic motivation for creating systems.[271] Of course, not all systems are made with the intention of combating a lack of honesty; Nietzsche, we must suppose, especially had legal systems in mind. But there are many other kinds of system. In a sense, food production is like a system, constructed not to combat a lack of honesty but to rationalize food production practices. And yet the setting up of this complicated food production systems also seems to be a good place for pushing honesty and integrity to one side. Today, consumers have no choice between personal trust and system trust; they are obliged towards the latter. Trust though is not the most appropriate word, as the food market in many cases is better characterized by a lack of trust, i.e. system distrust. Internal control and external monitoring are indispensable means to elevate system trust among users of the system. This control

[270] Giddens, Anthony: *The Consequences of Modernity*, pp. 17–21.

[271] Nietzsche, Friedrich: *Twilight of the Idols*, p 8: "I mistrust all systematists and avoid them. The will to system is a lack of integrity."

and monitoring must contain a strong element of distrust or suspicion – otherwise there would be no need for it. Systems and organizations always have some kind of internal monitoring of the actions of their employees. System distrust arises when the control and monitoring systems do not work properly, when they are not comprehensible for the consumers or not sufficiently independent and impartial in relation to the systems they monitor.

With regard to the food market, it seems questionable to make as sharp a distinction, as Luhmann does in his book on trust, between personal trust and system trust. True, he notices that mutual relations and dependencies inspire confidence, but he does not reflect further on the relationship between personal trust and system trust. These are not independent of each other, and in the case of food it is clear that system trust, or distrust, can be enhanced by personal contact with representatives of the food production companies. Giddens also refers to this with the concept of "access points", denoting points of connection between lay individuals or collectives and the representatives of abstract systems; "these are the meeting-ground of face-to-face work and of faceless commitments".[272] Thus, the style of the representatives is of the utmost importance for how the users experience the reliability of the system. Actually what happens at these access points is that system trust is replaced by personal trust – except that the representatives represent not only themselves but a system.

Personal contact is based on physicality and self-experience. Personal trust is different from system trust in respect of social ties, which for better or worse enhance the possibilities of exerting influence on each other. Luhmann writes that trust relationships find favourable soil in contexts where the participants know that they are bound to see each other again.[273] With Lévinas we can say that in the face of the other there is an ethical appeal to care for that other. Personal trust is based not only on prior information but also on the appearance of the other – on his or her style and character. And we may be good or bad at judging trustworthiness from these perceptions – which is of course the risk we run in any kind of trusting.

The direct contact between the participants, the face-to-face meeting, not only promotes trust relationships (compared to system trust) but also makes it easier to incorporate the expectations of others into one's own acts. It becomes easier for the producer or retailer of food to run their business in accordance with the expectations of consumers and conversely easier for the consumer to understand the conditions under which the production is taking place.

In conclusion we can say that on the food market personal and system trust go together. Normally, consumers cannot rely on just one of the two. Independent of each other the risk of misuse, the breach of trust, is huge.

[272] Giddens, Anthony: *The Consequences of Modernity*, p. 83.
[273] Luhmann, Niklas: *Trust. A Mechanism for the Reduction of Social Complexity*, pp. 36–37.

Nevertheless, on the food market most authorities and companies concentrate on system trust, which in many cases has not been successful enough to secure consumer trust.

2. FOOD AND ETHICAL IDENTITY

The consumption of foodstuffs presents itself as an opportunity for consumers to make an ethical choice in relation to man, society and nature. "Was ist der Mensch?" asked Kant. And Feuerbach answered: "Der Mensch ist, was er ißt". Feuerbach's statement makes no sense from a purely materialistic viewpoint; only as an aphorism can it have any meaning, in the sense that food and foodstuffs play a role in our own self-understanding. What is man? The question can be answered by referring to Simmel's sociological view of modernity: man is he who produces goods, exchanges goods and consumes them. This is of course far from an exhaustive definition, but it is not an insignificant one. The argument that food contributes to making man what he is rests on the belief that in a sense the food he eats *mediates* his relationship to himself, to society and to nature – it links him to himself and to his environment too. *You are what you eat* is of course an exaggeration, for there are many other things that contribute to self-understanding; one's self as a consumer is of course only part of oneself.

We saw in Chapter One that dietetics in ancient Greece was a particular way of constituting or creating oneself as a physical and moral subject. Food formed part of the technique of the self for creating itself as subject. The framework for dietetics as for ethics was closeness, the proximate, but today both must be extended to the long range, to the absent. The healthy body is no longer merely a question of what we eat but also of how it is produced – for instance, of whether chemicals are used in the production of vegetables. In a similar sense, the technique of the self for creating identity can no longer be limited to food and to physical health but must contain knowledge of, and an ethic for, the production history of the foodstuff in question.

The search for self-identity, for an answer to the question "Who am I?" has occupied philosophers since the very beginning of philosophy.[274] According to Heidegger, the understanding of one's own existence is the act of philosophizing itself.[275] Food can play an important role in this creation of identity and self-understanding, as it mediates a relationship to the environment, to other people and to society. And *can* is used precisely, for food *can* be of more or less importance for one's self-understanding. I have already mentioned different kinds of self-understandings formed by our relationship to food: how eating

[274] See, e.g. Martin, Luther H., Huck Gutman and Patrick H Hutton (eds.): *Technologies of the Self. A Seminar with Michel Foucault*.

[275] Arendt, Hannah: "What is Existenz Philosophy?" p. 48.

brings us into intimate contact with our surroundings, how the sharing of food is basic to communities, how food is used as a metaphor in Christianity and dietetics to show a care for oneself, but also how the symbolic value of food is reduced in the New Testament. It may have been noticed en route that these cultural meanings of food seem to be less concerned with the "self". And we may justifiably ask at this point: Are these really self-understandings? The answer depends very much on the definition of the concept of self. The word "self" tempts us to look for some isolated thing, the self itself. But a self taken in absolute isolation is meaningless. Rather, the self becomes a self in relation to something else. The self is formed in relation to others and in relations between human beings, which is often referred to as intersubjectivity, and also in relation to other things. Identity and the self do not exist independently of these relationships.

Man's identity cannot be regarded as coming from within, as innate; rather it is created in man's relation to his surroundings. The French writer and war pilot Antoine de Saint-Exupéry (1900–1944) was particularly interested in the relationships that man enters into, and how through these links he becomes part of a whole, which is different from the parts but which also contributes to his self-understanding. He describes man as a bundle of relations – *l'homme n'est qu'un nœud de relations* – and pushes this to extremes by saying that things in themselves are nothing, they only exist through their context. Relations cannot build on indifference, which creates no relation: they can only build on sympathy or antipathy.

Defining oneself negatively or oppositionally places one in contrast to others and the Other: e.g. I am not one of those who pray to God; I am not one of those poor old souls; I am not one of those who eat meat; I am not one of those who eat pork; I am not one of those who have a romantic nature; I am not one of those who lie; I am not one of those who believe in solidarity and so on. In the immediate sense, this says nothing about who I am, only about who I am not. But since such statements always exist in a context, they reach beyond themselves. If someone says they do not eat meat, we know that it is not because they are a sheep; we immediately localize that person as having a particular view of animals and possibly also certain observations regarding health and ecology. Saying who they are not allows us to infer or even understand who they are. On the other hand, this oppositional identity does not acquire its strength by that person through negation, allowing us to *see* who they are; confronted with the desire of post-modernism to deconstruct every belief and present all that we call truth as mere appearance and collective falsehood, it is difficult for us to find a convincing, unpretentious and non-superficial identity. Instead it is the oppositional in itself that provides identity.

The oppositional also builds on a power relation. Since power is exercised in a relation between the subject and the other, a distinction is drawn between the two that compels the individual into an identification process.[276] The importance of this for an oppositional, combative soul must not be underestimated. It is hard enough to know what to believe in a modern world where values are relative and deconstructed, and for this reason opposition as such can play an important role in the creation of identity. For the oppositional, confrontational person it is enough to say: "I am the one who won't . . . and I am against this or that". In *The Citadel* Saint-Exupéry has the first-person narrator, the Desert Prince, say that "by limiting you your enemy gives you form and constructs you",[277] and he even goes so far as to add that one could not come into being as a person without enemies! In other words, if we have no enemies, we make them. This also means that in the Desert Prince's philosophy we cannot live without our enemies and that we should therefore look after them: "And yet life is a network of such complicated relations that if you destroy one of the two antipoles, then you die . . . . The same is true of the one who destroys his enemy. Yet he lived by virtue of him. That is why he also dies with him".[278]

How can the oppositional identity assume such an important character that the Desert Prince can actually say that one is nothing if it is not present? "Identity" derives from the Latin *idem*, meaning "the same", but here we are dealing with an identity that appears to be based on our being different. The explanation is that the hostile picture makes us want to be divorced from the rest and thus it becomes possible to distinguish between "them" and "me". But this difference also means that "I" remain the same, the self-same, in opposition to my enemies.

The hostile picture is embedded in two things: first, as a relation between good and evil, similar to the bipolar relation between black and white. I am good and my enemy is bad. The good action cannot be understood without the idea of a bad action. Good and bad bear on one another and cannot be understood in isolation. The bad sculptors fertilize the earth for the good ones, says Saint-Exupéry. Understanding the good act requires a background of a bad or evil act. If my enemies are those who commit bad acts, I understand myself as one who performs good acts. This is a powerful identification: I am the one who does good, and I stand by it; that is my nature.

Political consumers acquire their self-understanding very much through this oppositional identity. As we saw in Chapter Three, consumers' political purchasing is not a choice *for*, but rather a choice *against*. They choose against the things they wish to avoid – when there is an alternative, that is. The boycott is defined by antipathy rather than sympathy.

[276] Dyrberg, Torben Bech: "Foucaults magtbegreb og dets relation til politik og demokrati", p. 123.
[277] Saint-Exupéry, Antoine de: *Citadellet*, p. 118.
[278] Ibid. pp. 246–247.

Against this, one could argue that ecological milk is a good example of how by opting *for,* rather than against, "green" consumers have had a decisive influence on production. But this is only a partial truth. For many consumers choose eco-milk, but what do they know about ecological production or the animals involved? Is it not rather the case that the choice of eco-milk has become primarily a choice *against* pesticides and fertilizers and less a choice *for* a particular production form? At present many producers of eco-milk are in difficult financial straits and a number have already closed because the prices they can get for their milk are too low in relation to their costs. The reason is that more eco-milk is produced than is consumed, so the price is lowered because supply outstrips demand. Producers must then work harder and rationalize, but by and large it is only they themselves and the dairies who realize this – and it is hardly a problem that political consumers are likely to focus on. The market will doubtless regulate itself. When a sufficient number have gone out of business and milk production has been further rationalized, a new balance will arise with lower prices and more efficient production. Political consumers cannot opt for or against the situation; they can continue to consume eco-milk, but however much they are willing, they cannot pay a price that secures the farmer a reasonable basis for production. What more can political consumers do? Not much, except to wait for the effects of rationalization, when the small farms have disappeared and the remaining ones have become even bigger.

The oppositional identity is negative and therefore unsatisfying in the long run. Instead, let us turn to MacIntyre and Ricœur to see what the concept of the narrative identity can mean in relation to consumers. In his book *After Virtue* MacIntyre writes that to understand what a virtue is, it is not enough just to understand how it may be exhibited in character, e.g. as honesty, reliability, courage and perseverance, for we only really understand a virtue when experiencing it, i.e. when it is enacted in a story.[279] "Man is a story-telling animal", says MacIntyre, and it is through the story about oneself that we find our own narrative identity. MacIntyre describes this identity as *the narrative concept of selfhood*; it is the identity that human beings acquire from their life story – the story about oneself. We are the subjects of our own story and yet at the same time we participate in the life story of others, just as they do in ours – the stories are interwoven. If by the virtues we mean the laudable qualities that can guide us through our life stories so that life is successful in an ethical sense, we must give credit to the consumers who purchase in accordance with their vision of the good life for and with others and not out of envy or economic self-interest based on cheapness. This is a virtue because on the market it is certainly not a matter of course to act in accordance with one's vision of the good life with and for others.

[279] MacIntyre, Alasdair: *After Virtue*, p. 125.

In Ricœur's *Time and Narrative* (Vol. III) we find the concept of *identité narrative*.[280] Ricœur refers to Hannah Arendt in stating that the answer to the question "who?" lies in telling a life story (French: *l'histoire d'une vie*). We explain who someone is by referring to their actions. However, the narrative identity is neither stable nor without cracks, for it is always possible to tell the story another way, for instance by including further or fewer elements, just as it is possible to see and compare one's own story in the light of fictive narratives. Thus the narrative identity is constantly being reconstructed since we are always evaluating our life story from the present point.

In *Oneself as Another* (1990) Ricœur makes the distinction between "sameness" (Latin: *idem*) and "selfhood" (Latin: *ipse*). *Idem* expresses the identical, a form of likeness, whereas *ipse* as a concept of identity expresses the permanence of a person despite the continuous changes of life. Peter Kemp writes that we learn about *idem* when we ask "*what* is the same" and we learn about *ipse* when we ask "*who* is the same".[281]

The narrative identity in *Oneself as Another* is played out between *idem* and *ipse*. In the narrative or the life story, says Ricœur, we find a link or dialectic between *idem* and *ipse*.[282] The narrative contains actions and actors (who may also be called characters). If we ask *what* is the same in a person (cf. *idem*), we can answer "their physicality", but we should also include their skills and character. If on the other hand, we ask *who* is the same (cf. *ipse*), we would not normally answer that it is the same person "who has that body" (making it sound as if there were two different bodies) or it is the same person "who has these particular abilities" (which could sound schizophrenic). Rather, we are inclined to define the person in question by referring to other means; it is one and the same person who has done this, or the same person who does such and such. If we ask *what*, we are asking within the boundaries of space, whereas the question of *who* relates to time. The dialectic between *idem* and *ipse* consists in the characters themselves deciding their actions while simultaneously improving or retaining their characters through performing them.

Returning to consumers, we can see that if we know the production history of the foodstuff in question, it will constitute part of, or become interwoven with, the consumers' own narrative identities. We could also say that such a production history is a narrative with which by virtue of being consumers we become interwoven. Knowledge of the production history explains something about the contexts into which consumers enter. This explanation offers a better understanding of the contexts we share with other people and with nature. We acquire a better self-understanding, because the role we play is part of a wider

[280] Ricœur, Paul: *Time and Narrative*, Vol. III, p. 246.
[281] Kemp, Peter: *Praktisk visdom*, p. 28.
[282] Ricœur, Paul: *Oneself as Another*, p. 140.

context; we get "into character", so to speak, which as we saw earlier links us to values and morality.[283]

Ricœur goes further by saying that there is a relation between narrative and ethics.[284] The narrative is never ethically neutral. In the narrative, experiences are exchanged between people concerning actions that are successful or unsuccessful. The ethical dimension of the narrative also reveals itself as a description of people with particular skills or virtues (cf. also MacIntyre above). The identity that the use of foodstuffs contributes to consumers, via their relation to the production history, is of course significant for the consumers' own self-understanding; it places it in a context: "I am the one who consumes foodstuffs which are produced in this way or that". And the manner in which they are produced relates to the people who are involved right back to the production organisms and "nature in the wild"; this therefore is "my way" of relating ethically to the other, to society and to nature. It would be a meagre identity if it stood alone, but the production history is merely one of many that are interwoven into our life story and contribute to our own narrative identity.

In not being immaterial the production history must mean something for us. The narrative understanding of a given foodstuff's production history brings awareness of a relation and an affiliation. What is significant for, and close to, us takes on an ethical character. Understanding the production history can be described as a refiguration (cf. Chapter Four). But the refiguration points not only back in time but also forward to the future as a *horizon of expectation* (German: *Erwartungshorizont*),[285] as an expectation of how the future will look based on past experiences. The horizon of expectation is more than this, however, for it may also be said to contain the idea of how the future *could* be. When we hear or know the production history, we will be transported into the horizon of expectation, and on the basis of our experience with how things are, we can reflect on how things could be, be they better or worse. Through the horizon of expectation knowledge of the production history can become a vision of a fair food production for and with others.

3. FOOD AS A TRACE OF NATURE

We have already seen how foodstuffs are being linked less and less to nature and otherness. They are being subjected to a pre-bodily digestion that persuades us to see food as culture and as something already becoming ourselves. Through the aestheticizing of the meal the nature element in food is moving into the background.

[283] Ricœur puts it thus, "expliquer plus pour comprendre mieux". Through *explanation* we acquire greater *understanding*.

[284] Ricœur, Paul: *Oneself as Another*, pp. 163–168. This is already more than is hinted at in Chapter Four.

[285] The concept of a *horizon of expectation* is found in Reinhart Koselleck: *Futures Past. On the Semantics of Historical Time*, p. 267.

This distancing from nature is typically expressed in three different ways. First, nature is experienced as being insignificant, since it plays little or no role for modern man, partly because it has no *visible* place in modern society. Second, nature may be experienced as a threat to culture and human society, in which case it must be subjected and limited. Third, nature may be regarded solely from its usefulness. Utility becomes the only logical approach to nature and thus the only value to relate to – a view found, for example, in Marx – where man's relation to nature in particular is instrumental and technical. On the relation between the view of nature and the application of technology Marx writes:

> Technology discloses man's mode of dealing with nature, the process of production by which he sustains his life, and thereby also lays bare the mode of formation of his social relations, and of the mental conceptions that flow from them.[286]

According to Marx, industrialized man regards nature as raw material and therefore treats it as such. An industrialized society is very much an urbanized society, where practically everything is created for a purpose and serves some function or other in relation to man's life. When urban man meets things that are not created for a purpose, he can be seized with wonder. Bertolt Brecht writes that "in time it confuses us only to see articles for everyday use in the city, in houses, on the railways, which if they were uninhabited or unused would be meaningless".[287] Urban man therefore tends to understand nature in terms of utilization, function and purpose; and what is useless worries him, even threatens him – he does not wish to know.

Marx was primarily interested in the meaning of nature for human activity and he therefore believed that nature in itself could be nothing for man.[288] Here we should note the preposition *for*, since in his theory of knowledge Marx retains external nature and its laws as a reality,[289] but allows man no immediate access to it, since it is mediated by his social condition. This involves more than just the glasses through which one views nature; for Marx it is a quality of the eyes that see. Nature is mediated to man through *society*. There is no pure, historically unmodified nature as a cognitive object for the natural sciences. All nature only becomes meaningful in the historical framework of the social processes. As far as human labour can stretch, it changes nature-for-itself into nature-for-us.[290]

Any view of nature can be regarded as an ideology. Marx speaks of the ideological character in the pathos of nature. In the instrumental view the care of

[286] Marx, Karl: *Capital*. Pt IV, Production of Relative Surplus-values, ch. XV, Machinery and Modern Industry, n. 4.

[287] Brecht, Bertolt: *Geschichten vom Herrn Keuner*. Quoted from Seel, Martin: *Eine Ästhetik der Natur*, p. 105.

[288] Alfred Schmidt: *The Concept of Nature in Marx*, p. 30.

[289] Ibid. p. 33.

[290] Ibid. pp. 49 and 76.

nature is legitimized by self-interest in maintaining the means of production for the future, a narrow economic rationality governing the relationship. And if we follow Max Horkheimer (1895–1973) and Theodor Adorno (1903–1969) in their book *Dialektik der Aufklärung* (1947) there is little hope for the future:

> Precisely because the social mechanism of domination seizes onto nature as a wholesome contrast to society, nature itself is bartered away and rendered unwholesome.[291]

In viewing nature we will always be ruled by our prejudice and ideology, and it makes no sense to speak of any genuine or original relationship to it; all such references are deception and seduction.

Even in *Worpswede* (1902), which describes five artists in a colony in Worpswede (in north-west Germany), by the otherwise romantic German writer Rainer Maria Rilke (1875–1926), we find a fundamental distinction between man and nature. Rilke's acquaintance with Worpswede lasted only a few years before he became secretary to the French sculptor Auguste Rodin. Within a year of its publication Rilke was already distancing himself from the book, calling it "invalid". The description of the paintings is of no particular interest, but in the introduction Rilke makes some general comments on "painting a landscape". In the situation where the artist is facing a landscape and nature, and must transform his sense impressions of it onto canvas, he is forced to reflect on the relation between man and nature, for what is one to paint? Should it be the most correct reproduction of what the eye sees, the faithful copy of nature, where the personal interpretation is minimized? Indeed is that at all possible? Or should one allow personal experience to dominate? Should one focus on the division or the harmony between man and nature? Artists must decide what it is they see, what it means to see and what they wish to capture in the painting.

For the Rilke of *Worpswede*, whoever is to tell the story of the landscape must abandon themselves to the foreign, the unknown and the unfathomable. Man is used to meeting and dealing with the will of another, but even though the landscape moves, the trees grow and the water babbles, nature does not *will* anything, it has no purpose:

> Let us simply admit it: the landscape is foreign to us, and we are frightfully alone with blossoming trees and babbling brooks. Alone with a deceased person we are not nearly so abandoned as when we are alone with the trees. For however secretive death may be, so much more is a life that is not our life and which does not participate in our life, and which, without seeing us celebrate its festivals, we can view with a certain self-consciousness as we do casually passing visitors who speak another language.[292]

[291] Horkheimer, Max and Theodor W. Adorno: *Dialektik der Aufklärung*, p. 177.

[292] Rilke, Rainer Maria: *Worpswede. Monographie einer Landschaft und ihrer Maler*, p. 9: "Denn gestehen wir es nur: die Landschaft ist ein Fremdes für uns, und man ist furchtbar allein unter Bäumen, die blühen, und an Bächen, die vorübergehen. Allein mit einem toten Menschen, ist man lange

This understanding of nature must be called pessimistic! Although Rilke remarks that nature celebrates its festivals, it remains secretive to him. Nature does not care whether we are present or absent, for it seems as if nature does not notice our presence. Highways, byways, agriculture and the city also pass unnoticed by nature. It matters not to nature that we dig for iron and oil or that we till the earth and reap its fruits; nature does not participate in our lives, our joys and sorrows. Nature just does not care; it carries on as usual, which is all it *can* do. We cannot make ourselves understood to it and it remains dumb to us; indeed it means less and less to us now.

But it is also a one-sided view to see nature solely for its utility value and its strangeness. The English poet-painter William Blake (1757–1827) wrote of man's twofold attitude to nature in a letter from 1799:

> I feel that man may be happy In This World, And I know that This World is a World of Imagination & Vision. I see Every thing I paint in This World, but Every body does not see alike. To the Eyes of a Miser a Guinea is far more beautiful than the sun & a bag worn with the use of money has more beautiful proportions than a Vine filled with Grapes. The tree which moves some to tears of joy is in the eyes of some others only a Green thing that stands in the way. Some See Nature all Ridicule & deformity, & by these I shall not regulate my proportions; And Some Scarce see Nature at all. But to the Eyes of the Man of Imagination, Nature is imagination itself. As a man is, So he Sees. As the eye is formed, such are its powers.[293]

For Blake the modern man who sees only "ratio" or reason is blind to the nature of the mind. "As a man is, so he sees", he writes. Marx would perhaps say that Blake accepts that our view of nature is conceit and fantasy, but for Blake it was less a question of finding an original and genuine understanding of nature than of nature providing an inspiration for the mind, and thus it matters little whether it is true or false, for there is no ultimately true understanding of the world or of nature – we must create that ourselves.

The American writer Henry David Thoreau (1817–1862) sees sensuality as having been repressed by civilization and quarantined so that society can function and man can work and fulfil his social commitments. In this sense man is bound to society, tamed by work and duty. In his meeting with nature it becomes possible to forget these wearisome matters and return to sensuality. The reconquest of sensuality is liberating:

> Of course it is of no use to direct our steps to the woods, if they do not carry us thither. I am alarmed when it happens that I have walked a mile into the woods bodily, without getting there in spirit. In my afternoon walk I would fain forget all my morning occupations and my obligations to Society. But it

---

nicht so preisgegeben wie allein mit Bäumen. Denn so geheimnisvoll der Tod sein mag, so geheimnisvoll ist ein Leben, das nicht unser Leben ist, das nicht an uns teilnimmt und, gleichsam ohne uns zu sehen, seine Feste feiert, denen wir mit einer gewissen Verlegenheit, wie zufällig kommende Gäste, die eine andere Sprache sprechen, zusehen".

[293] Raine, Kathleen: "William Blake – Prophet of Imagination", pp. 6–9.

> sometimes happens that I cannot easily shake off the village. The thought of some work will run in my head and I am not where my body is – I am out of my senses. In my walks I would fain return to my senses. What business have I in the woods, if I am thinking of something out of the woods? I suspect myself, and cannot help a shudder when I find myself so implicated even in what are called good works – for this may sometimes happen.[294]

Marx wrote that we see nature through society, that the state of society is simply an attribute of eyes that see; but for Thoreau, who is also aware of this view of nature, as when he takes his duties to society out with him on his morning walk, it is also possible to shake it off and forget it in order to experience nature in a way other than the utilitarian.

In his book *Eine Ästhetik der Natur* (1991) the German philosopher Martin Seel (b. 1954) describes three aesthetic ways of relating to nature, which he also contrasts with the instrumental view of nature. His fundamental hypothesis is that it is impossible to remain neutral in relation to nature; as with Kant, only the individuals can set the yardstick for their view of nature. In Seel it is both *Vorbild* and *Nachbild* – our model, yet something we ourselves create – thereby transforming the question into *the right relation between our ways of relating to nature*.[295] There are many ways of doing so, meaning that the relationship relates in fact to itself. The aesthetic understanding of nature is also a construction or a relation through our relating, as Seel writes.[296] He understands nature's aesthetic to be part of the ethical theory of the nature that constitutes man's life world (in contrast to nature in the natural sciences), precisely because it illuminates the various ways of relating to nature.[297]

Seel distances himself from the idea that nature should only be regarded in a single way, preferring various ways but more or less simultaneously. In addition to the philosophy of its instrumental usefulness, which Kant links to self-interest, nature as an aesthetic can also be viewed through (1) *contemplation*; (2) *correspondence*; and (3) *imagination*. In Seel's view, the reason why nature pleases us and therefore belongs with aesthetics is that it is vulnerable and corruptible (*zerstörbar* in the sense of "perishable"), and that our relation to what can be destroyed refers to a practical problem that reveals itself in the life world and not in a theoretical and objectivable world.[298]

The contemplative approach that we saw in Thoreau is to turn away from life's daily chores and regard nature with disinterest: the lake at Walden is not a symbol of anything, nor is it classified as a particular type of lake; it is only viewed as it appears, phenomenologically, before we begin to grasp it with the intellect and reason. Nature is "foreign" in meaning (*sinnfremd*) because it does

[294] Thoreau, Henry David: *Walking*.
[295] Seel, Martin: *Eine Ästhetik der Natur*, pp. 14–15: "The question of the proper relationship to nature becomes the question of the proper relation between our ways of relating to nature."
[296] Ibid. p. 35.
[297] Ibid. p. 31.
[298] Ibid. p. 22.

not reveal any meaning or function for us. Thus, the contemplative view also dissolves or suspends the centredness of the ego.[299] The contemplative judgement is eccentric, says Seel. Here everything is without significance, yet everything counts. No matter what the contemplative attention directs itself towards, it appears beautiful because everything is present for the observation, leaving no room for understanding. All the senses, even smell and taste, can form part of this contemplative view of nature, but sight is superior in being able to distance itself from the object in focus. The exercise in the contemplative view consists of not only, and no longer, understanding things discursively.

The corresponding view is an existential observation of nature as a place of life: "To experience nature as beautiful in a corresponding sense is to see it as an expression of, and part of, the opportunity for a good life".[300] Conversely, ugly nature is a place without the possibility of a good life. In the positive correspondence, writes Seel, beautiful nature is transformed into part of one's own life; here is a place where I can live. In this positive correspondence we are attracted by nature and it is viewed as a possible part of our own life situation, not because nature has a meaning, but because we *find* a meaning in nature; we see nature in the meaning that it can acquire for us. The corresponding judgement is normative in so far as it says something about the idea of the life that is successful.[301]

Seel calls his third aesthetic understanding of nature its "imaginative" interpretation. Here we can regard nature *as though* it were art, well knowing that it is not – hence the term *imaginative*. But it is not just us projecting our artistic imaginations onto nature, for nature also helps to establish the framework around which we can have any imaginative ideas at all. Nature promotes fantasy.

The consumption of foodstuffs mediates a particular handling of nature, because they are produced through farming, which is related to nature practically as a way of "treating" it. "Farming symbolises our space", says the Japanese anthropologist Emiko Ohnuki-Tierney (b. 1934),[302] and points out that food and foodstuffs have a close relation to the animals and plants that are found in the universal landscape: indeed foodstuffs actually *are* the animals and plants that we find in the landscape. The landscape and nature are dominated by the cultivated soil and thus a relation to food and foodstuffs, whether conscious or not, is a relation to nature.

Market and economic rationality and science safeguard the instrumental view of nature, where the production must be effective and costs minimal. The dominant conception of domestic animals within modern rational agriculture is that

[299] Seel, Martin: *Eine Ästhetik der Natur*, pp. 14–15: "The question of the proper relationship to nature becomes the question of the proper relation between our ways of relating to nature." p. 57.
[300] Ibid. p. 90.
[301] Ibid. p. 133.
[302] Ohnuki-Tierney, Emiko: "We Eat Each Other's Food to Nourish our Body: The Global and the Local as Mutually Constituent Forces", p. 245.

animals and plants exist for *us*. They are assessed for their usefulness to us humans. And perhaps that is not so strange, for it is the only meaning they have in industrialized farming. The animals are isolated and instrumentalized in the production halls of modern agriculture and have no other function than to become foodstuffs, cut off as they are from other contexts. If as consumers we believe that it is sufficient for food ethics to settle for adopting an instrumental relation to nature, we can safely trust the market. But if through our consumption of foodstuffs we wish to show other considerations for nature, among which is to be found in the aesthetic and the environmental view, then the market is less well suited. Against this, one could argue that aesthetics have nothing to do with farming; that kind of experience belongs to another nature where there is no agriculture. This is also the case for those consumers who can settle elsewhere or move to areas away from agricultural production. But farming is an important part of the landscape in which people live, and it *is* the daily space for many people; it is not therefore asking too much to insist that food production should pay regard to the aesthetic experience of nature. For as Seel writes, the aesthetic experience of nature then becomes part of the good life that succeeds.[303]

We have already noted that we cannot know whether nature has any intention or meaning in itself and it is therefore difficult to argue that we should pay regard to nature for nature's own sake. "The value of life cannot be assessed", says Nietzsche.[304] Seel observes that nature assumes the value that we ascribe to it, and that it acquires its value through either the engaged or the distanced view.[305] Meaning is not just something that comes to us from without, but something we ourselves help to create. This is an important point, for as Nietzsche also writes, then people do not suffer through their actual suffering but through the meaninglessness of that suffering.[306] Unless we ourselves are able to invest life with meaning, it will doubtless appear meaningless.

Seel emphasizes the variety and otherness of nature as significant elements for the aesthetic experience of nature. Nature's otherness according to him can be described with the concepts of independence (though not in the autonomous sense of "self-regulating", but as independent in relation to man) or virginity, but Seel is also aware that when applied to nature these concepts are not without problems, since already in man experiencing nature, that virginity is being invaded. Seel nevertheless insists that the independent, unforced and unordered nature is the model for an aesthetic relation to nature.

Food consumption is a way of relating to nature, and I have described two contrasting views on nature that consumers may wish to take into consideration:

303 For Seel the natural beauty in the three aesthetic experiences of nature is linked to the good life that succeeds – and thus to ethics. Seel, Martin: *Eine Ästhetik der Natur*, p. 290.

304 Nietzsche, Friedrich: *Twilight of the Idols, or, How to Philosophize with a Hammer*, p. 12.

305 Seel, Martin: *Eine Ästhetik der Natur*, p. 314.

306 Nietzsche, Friedrich: *On the Genealogy of Morality*.

the instrumental and the aesthetic. Omitted from this description is the issue of environmental sustainability, which of course is also of utmost importance for many consumers. However, including the instrumental and the aesthetic view on nature, it becomes clearer that food production practices form the landscape and nature and that this is an issue of ethical concern. It also becomes clear that there are many and complicated elements for the ethically concerned consumer to take a stand on – and that these are often negotiated within the consumer.

4. FOOD ETHICS OF THE CONSUMERS

Food ethics are undergoing considerable changes in these years. Having been excluded from debate for decades in the wake of the industrialization of food production, they are now attracting greater interest among producers and consumers alike. There is a growing feeling that severing the link so sharply between food and ethics was perhaps unwise. The search for a new understanding of food ethics involves two elements. First, the new and somewhat vaguely termed "globalization". The world has "opened itself" to the extent that we can now have relations to people all over the globe. Conversely, it has also become smaller, and what once seemed far away is now close at hand thanks to modern communications and transport. This is significant, because our actions now have not only local but also global consequences. Second, the historical aspect is back on the agenda, i.e. an interest in how people in the past related to food ethically. Perhaps we can learn or rediscover something by studying historical sources pertaining to food. In simplified form we might say that before we became "modern", food ethics was understood in two different ways: individually and socially.

Individually, as a *technique of the self*, it was held that the choice of food expressed a relation to the self and its surroundings. Food created identity and had an ethical content, for it had to do with the good life and one's own body. Most people still relate to food in this way, whether consciously or unconsciously, but in certain philosophers this attitude is pronounced. The ancient cynic philosopher Diogenes (*fl.* 4th century BCE) regarded man as a biological and animal being, and he therefore shunned prepared food in favour of eating like the animals – only raw meat. The Romantic French philosopher Jean-Jacques Rousseau (1712–1778) preferred simple country fare with lots of milk and bread rather than elegant gastronomic dishes. Both regarded food as an important part of their self-understanding, as an extension of their philosophy and ethics, so to speak. At the same time their culinary habits constituted a form of cultural criticism. In their food choices both rejected "civilized" society. Diogenes wished to live in an original state of nature as an animal, while for Rousseau the healthy peasant culture was the ideal life form.

The *social* meaning can be explained through the usage of the German word *Gericht* (Scandinavian *ret*), which has both a gastronomic and a legal meaning (where English has *dish* and *right*). For the community the basic *Gericht* can be interpreted as the right to a dish, meaning the right to share the food available to the community. Food and rights (dishes) are thus a way of expressing a care for the other person and for the community.

Food ethics in the past may have seemed relatively simple and concrete, but that is far from being the case today. In modern society, production has moved outside the home and the local community, making food ethics no longer a technique of the self and a right to food, of *having something to eat* and of *what one eats*. Today it is just as much a question of *how the food one eats is produced.* In fact, we see the first inklings of this understanding of food ethics already in Rousseau. In *Emile, or On Education* (1762) Rousseau gives prominence to the country life as the original and the good life form. The basic knowledge of cultivating the soil, i.e. *how* to produce food, is central to Rousseau's understanding of the good life. But his "food ethics" are premodern and romantic in regarding the division between production and consumption as a split.

If we agree that a return to a more or less original rural life where we are all part of the food chain is not a good idea, and is anyway impossible, then food ethics must rest on another foundation. Since most people have no direct contact with food production, the first task of food ethics is to examine how to communicate the essential knowledge of the production history. The intellectualization of foodstuffs means that we must reconsider the role that they play for the individual, for society and for nature. This tour around the world of food and philosophy is one such attempt to find a new fulcrum for food ethics.

Already at the outset, however, comes the first confusion. Food is a multifarious subject, with ramifications for large areas of society. With so many balls in the air at the same time, it is easy to drop them all. For the inevitable question will be asked: What does this or that mean for food ethics? What does the cultivated distancing to food or the political consumer's impotence mean in relation to food ethics? Or what does the ethic of closeness as developed here actually involve, and how can it influence the consumers' own shopping?

At this point in the book the reader is hopefully wiser about the arguments, but what are the practical results of this study? For there are no clear instructions as to how the thoughtful consumer should act, and few suggestions as to how new approaches can be implemented in practice. The ethic presented is barely "tangible" in the sense of explicit values or norms for the right action. There are no lists of products to be placed in the shopping basket with a clear conscience, and no overview of production conditions to be noted. There are no pictures of battery hens or happy free-range pigs, no bankrupted farmers or satisfied consumers or instructions for ethical labelling. In fact as a subject food ethics is here presented as universal and in the abstract. Neither the criticism of the intellectualized foodstuffs regime nor the developed food ethic as *the vision of the good life with and for others in a fair food production practices* contains simple concrete directions. So is it indeed an "ethics" that has been developed in these pages?

By way of answer let us begin by distinguishing between ethics, morality and norms. Norms indicate the right actions in the form of prohibition or injunction. Morality states the values that underpin those norms and thus those actions. These values can be justice, equality, autonomy, utility or they can be virtues such as courage, honesty, solidarity or responsibility. Ethics goes even deeper by seeking to justify moral values. This leaves them always subject to a kind of open ending, for in the ethical debate one must always be prepared to resume the discussion on the justification of values and to remain open to the possibility of revision. Ethics must never resort to dogmatic finger-wagging or moralizing political correctness. Ethics as a *vision* cannot be formulated in principles, rules or values even though these are necessary and part of the picture; ethics gives primarily meaning as a *view*, or a *way of viewing*. It is a view of what constitutes the good life and of models for action.

The good life does not stand alone; it always exists in relation to something else, to nature, society and other people. And whether we like it or not, our use of foodstuffs and food has a practical effect on nature, society and ourselves. When we eat, we are supporting, however unconsciously, a particular way of producing foodstuffs that has consequences for practical life. *Nature*, from where the foodstuffs derive, can be regarded from a utilitarian viewpoint or as a subject for immersion and empathy. Agricultural production can reflect these views in various degrees. *Society* is organized in different ways and with various methods of producing foodstuffs – e.g. a globalized food market involves much transportation. *Man* understands himself partly through the way his food is produced: "I am the one who eats food produced in this or that way". Most people actually realize that they are participating in industrialized agriculture and this is part of their self-understanding. However, this modern self-understanding implies that they do not know much about the production history of their food: "I am the one who eats industrialized food that I know nothing about".

Instead of trying to present correct solutions, the purpose of this study has been to enable readers (i.e. consumers) to draw their own conclusions in actual situations and thus develop their own food ethic; the book merely presents the conditions under which consumers may exercise that ethic). To a certain extent it is a subjective view, as it describes the conditions necessary for the individual consumers to make an ethical choice, but that does not preclude a widespread agreement on the need to debate a subject that is encroaching increasingly on our everyday lives. The demands on consumers are considerable. And it requires a particular virtue to be an "ethical" consumer, namely courage – the courage to act in accordance with one's own ethics. Unfortunately, ethical visions often face difficult conditions on the food market, because they do not harmonize with short-term economic interests. It costs time and effort and even money to have an ethic. And it does not make it easier that the cultivated distancing to the natural origin of food reflects a will to remain ignorant of food production practices. But primarily it is difficult because the impenetrability of the food market means that the consequences of the consumers' purchasing remain hidden.

If food ethics is to plead its case in modern food production, consumers must have better knowledge to understand the effects of their consumption. This is not an easy task; one solution put forward here is to mediate to them the *production history* of a given foodstuff, not as branding and marketing where more or less "virtual experiences" and fictive stories are told, but the actual true history of the origin and making of the foodstuff. Food ethics then becomes a narrative ethic, a way of viewing the true history. The narrative form is seen as fundamental to our being at all able to understand what the good and the bad actions are. Presented as narrative the production process offers consumers the opportunity to make a critical response: could the production history be told differently? Is it good enough as it is or could it be made better?

The production history in itself is of course not sufficient for ethical engagement. Closeness to the product can motivate this through sensuality, physicality and the existential conditions they involve, because we better understand the effects of our actions as they become tangible. To achieve this closeness, the trace of a foodstuff is a valuable element. Consumer food ethics must therefore build on both sensuality and intellectual knowledge.

A central question that the reader has doubtless considered several times is why not leave food ethics to the authorities? Can the politicians not ensure that the ethics of the production history are sound? Perhaps it is not even in the consumer choice that decisions on, for example, production conditions should be made. The question is fundamental: Who makes decisions on food ethics and by what process? It is surely too much to expect busy consumers to become involved in complex cases surrounding modern food production.

In answer, we must of course expect public authorities to regulate via rules, standards and control checks and through public debate on what is or is not acceptable. On the free market there will always be producers who try to cheat, and they must be checked, for this the consumers cannot do. A regulation for a minimum of information on a given article is absolutely essential, as are rules on securing the working environment, animal welfare and so on.

However, the reason why the consumers themselves should make up their own mind is that the process has an effect on their self-understanding and their dignity: there is a world of difference between knowing and not knowing the effects of one's actions. In social terms, the consumers' wish to make ethical choices testifies to a shift on the political stage to also include the market.

This politicization of the market is often criticized by politicians and economists, who prefer to keep the two domains apart and within their current balance of power. Nevertheless, consumer interest in ethics and politics must be seen as a new development of democracy, though one not without problems. There is a clear need to interpret the political and democratic rules of the game in new ways and fields, consumption being one of them. Ethical consumption is a refreshing new challenge to democracy – and a timely one too.

The challenges are legion. Taking the political consumer seriously presupposes genuine information and genuine opportunity for action. The transparency and traceability that "ethical" consumers are requesting are also linked to the economy. Economic transparency is important for the consumers to be able to know *to whom* and *to what* they are paying. At the same time the amount of information on the food market is already overwhelming – and yet consumers cannot use large areas of it because they either do not understand it or they are not interested in it. What is needed is more information *of the right kind*. Consumers are reasonably informed on product contents but most certainly not on production methods.

The Nordic Council of Ministers has stated that "ethical consumption is diffuse for the consumers and is not an evident preference in the consumers' everyday choice of consumer goods".[307] Most of us would agree with this; we can *think* about the ethical implications of consuming goods, but it is seldom that we possess the knowledge or the access to information about the production practices to actually make an informed appraisal of the ethical implications of our choices. If consumers find ethical consumption to be diffuse, this is probably not due to their inability to judge what is the good life with and for others in fair food production practices. It is more likely due to a glaring lack of access to information. My personal experience with running a consumer-supported farm bears out that people very often possess an intuitive understanding, indeed a very clear one, of what is good and bad when presented with different food production practices, but not always of course.

What then is the hope for consumers' food ethics in the future? The answer is: their moral integrity. This may sound a little heavy-handed, but it refers directly to the consumers' will to see beyond, and then to move beyond, their own narrow self-interest – beyond the cheapness that this involves and into the production history that presents its own effects in a wider context. They will then see how the production history affects society and nature as well as their own self-understanding. The concept of "integrity" is normally used about people who possess the specific virtues of constancy and resolution.[308] Logically speaking, these qualities can be applied both to the saint and the sinner, to the

[307] Nordic Council of Ministers: *Forbrugernes fornemmelse for etik*, p. 9.
[308] MacIntyre, Alasdair: *After Virtue*, p. 203.

criminal who denies wrongdoing and to the honest man who accepts full responsibility. But denial of wrongdoing is not a virtue, and constant denial of one's bad acts would be absurd; hence denial cannot be "constant" in all the fields of life. Denial is never stable, but is inconstant and irresolute. Constancy and resolution can therefore be linked to "integrity" among those who do not deny their words and deeds but stand firmly by them:

> To be 'answerable' means that to the question of who committed this act one can *answer*: 'I did' or 'It's me'.[309]

To be answerable for our consumption demands not only that we say: "It's me who has consumed it". We must also be answerable for the effects of that consumption – and it requires resolution to act on one's vision of the good life.

The effects of consumption must not only be sought in the present but, as we have seen, in the past and the future also. The consumers' moral integrity therefore requires *a will to know* about the production history and to act (also commercially) in accordance with their ethical convictions. The production history of the food is important in many ways, not least because giving something a history is tantamount to ascribing a *meaning* to it.

[309] Kemp, Peter: *L'irremplaçable. Une Éthique de la technologie*, p. 44.

CHAPTER SIX

TRACEABILITY AND FOOD ETHICS

So far the food ethics developed in this book is on a theoretical and abstract level. In this last chapter I intend to give more clear and concise instructions as to the practical implications of the food ethics developed, i.e. how to act from a food ethical perspective. In other words, I will show that the theoretical reflections can be more than just an academic interest. They can be useful for people who are in favour of ethical consumption and as a means of political participation in the development of society.

One thing should be clear from the theory developed so far. From the consumers' perspective, food ethics relates to the production history of the food. This is so because it is in the production process that we find a certain vulnerability: human beings, animals and nature in general, which are all part of the production process, can be harmed and violated by specific production practices. This vulnerability makes explicit the ethical relevance of the production history. In cases where there is no vulnerability, it does not make sense to talk about ethics. As I have shown in the previous chapters, in the actual act of shopping or consumption food is no longer "living" and as such is no longer vulnerable. Hence it makes no sense to have an ethical concern for the actual vegetables in the shop or the steak on the plate. To make any sense our concern must be directed towards the life that the vegetables or the steak were once "part" of. Such life – the food that once "was" – is found in the production history.

The effort needed on behalf of the consumers seems intimidating and immense; tracking the production history is a difficult, if not impossible, task in a highly industrialized society. For instance, it is often impossible to trace the origin of cereals used for bread because the flour is a blend of cereals from many different farms mixed in huge containers at the mills. Cereals used in industrially produced bread are likely to come from several different countries, even from all over the world. Even dedicated consumers, for whom ethics have become an integral part of the shopping culture, in many cases have to rely on labels indicating specific production practices like fair trade and organic farming without any concrete knowledge of the actual production history. Production histories are simply not available to the normal consumer.

Hence, the food ethics presented in this book may seem of poor use to most ordinary citizens, simply because food does not function as a sign that reveals the actual production history, but only as a trace that tells us that there has been a production history. I would therefore like to conclude the book by trying to make the philosophy and ethics of the trace in relation to food more concrete and applicable. For this purpose I shall list four steps.

The first is to introduce the concept of traceability. Traceability has become a rather technical term designating the ability to trace and follow a product through all stages of production, processing and distribution. Thus, traceability implies that the history of the entity consists of recorded identifications. This makes it possible to identify all the ingredients of a food product. Keeping record of a product implies knowing who the supplier is, what is supplied, when it is received, to whom it is delivered and so forth. The concept has already been used for some time within the industrial production of medicine. Should a medical product turn out to be malfunctioning or to contain unwanted elements hazardous to health, it is important to be able to trace it and recall it. As a result of food scandals in the last decade traceability has likewise been introduced in the food sector by means of EU directives.

In the second step I will present the empirical results of a survey on traceability and ethics conducted in Denmark. As part of the survey we implemented traceability on selected products in six Danish supermarkets with special attention to ethical issues. The results and how the consumers reacted to the "experiment" will be presented.

In the third and fourth steps I shall discuss these empirical results with reference to two concepts that relate fundamentally to ethical consumption and food ethics, namely autonomy and recognition. These are basic ideas and requirements for people who wish to take an ethical stand in their consumption of food.

1. FRAGMENTATION AND TRACEABILITY

Modernity has often been characterized by the concept of *fragmentation*. Referring to the imagined closeness of the past, many would say that we human beings are no longer linked to each other in a visible and comprehensible manner. Rather, existence in modernity seems to "suffer" from disconnectedness and fragmentation. The separation of consumption from the production of food as described in Chapter One follows this interpretation of modernity, as in most cases no immediate and visible connection between people is possible. Fragmentation has become such a widespread experience for modern man, and hence a fundamental human condition, that the Greek philosopher Cornelius Castoriadis (1922–1997) considered it to be one of the most pressing problems of philosophy today:

> The world – not only ours – is fragmented. Yet it does not fall to pieces. To reflect upon this situation seem to me to be one of the primary tasks of philosophy today.[310]

The world is fragmented – but even so it does not fall apart. Castoriadis calls for reflection upon this. One possible interpretation is that we experience the world as fragmented because we are unable to see the links that keep the world together, not because we cannot see but because the links are invisible. This interpretation of fragmentation makes knowledge a central concept; we do not see the links because we do not know of them.

Returning to Castoriadis for another interpretation, he would not agree that it is only in modernity that we encounter fragmentation. Rather, the fragmented nature of the world is a fundamental character of the perception. Thus, Castoriadis makes fragmentation an epistemological question. Fragmentation is a way of organizing the world, e.g. in perception, where some spheres are excluded and others included. The world must be reduced and fragmented in order for us to create a comprehensible order. If not, perception would be an embarrassment with its innumerable details. We could say that the perception of the world is fragmented and precisely because of that our perception of the world does not fall to pieces.

Even so, modernity has contributed not with an epistemological but with a social experience of fragmentation, referred to in the first interpretation presented above. Fragmentation has become a common, existential experience; not only the world but also our lives are experienced as fragmented. Fragmentation is experienced as a lack of coherence in people's lives, especially in the form of dividing and separating the different spheres of the human life: personal relationships, family, work, leisure, consumption and so forth. Often these areas are distinct and without connections. But also within these groups fragmentation appears; we have personal relations with different people who have no connections with each other – indeed we might be part of more than one family. Work and consumption tend to be separated from the consequences of these activities.

The existential experience of the fragmentation of social life has consequences for our attitudes to the meaning and purpose of society and in general for what it means to live together with other people. The Canadian communitarian philosopher Charles Taylor remarks that people "may indeed feel linked in common projects with some others, but these come more to be partial groupings rather than the whole society".[311] People tend to give up carrying out projects of common interest or even showing solidarity and care for each other. According to Taylor the consequence of this is that "the absence of

[310] Castoriadis, Cornelius: *Le monde morcelé* (1990). The quotation here is from the foreword to *World in Fragments*, p. xxii.

[311] Taylor, Charles: *The Ethics of Authenticity*, p. 113.

effective common action throws people back on themselves".[312] This individualism in turn leads to a "responsibilization" of the individual. However, not in the sense that people necessarily act more responsibly; some people sink lower just as well as some people rise higher.[313] In the end, and this is Taylor's pessimistic diagnosis of modernity, this means that *solidarity* tends to be restricted to proximate settings with face-to-face interaction. Society as a whole is dominated by an ongoing liberalism of neutrality and is therefore much better characterized by *tolerance* than by solidarity. Solidarity is used to characterize relations that benefit or are enriched by the presence of others as in a community, while tolerance refers to a society where the presence of the other is merely accepted.

The perception of the fragmentation of modernity varies a lot. Some consider the inherent fragmentation of modernity a minor problem because the advantages and possibilities of modernity more than compensate for the losses. For instance, increased individual freedom might be preferred over the feeling of community coherence and embeddedness. But if we leave out considerations about benefits and disadvantages, I believe that most people would agree that fragmentation at some level also implies a loss. If fragmentation is inherent to modernity, and coherence and unity do not appear by themselves automatically, then it has become a task of human beings in modernity to create coherence in their lives. For as fragmentation is dominated by a feeling of meaninglessness and isolation, coherence makes meaning by establishing relations between human beings (as described in Chapter Five and as I shall return to in the last part of this chapter).

Fragmentation has also taken place within food consumption, in the sense of isolation from production, since the 17th century. Unity and coherence from farm to fork do not appear by themselves in modernity, they must be created. Production histories of food represent the link from farm to fork but they are not accessible to ordinary people. However, they can be *made* accessible. This is not an easy task for consumers but it can be done by those who are in possession of the information concerning production practices. Here of course I am referring to the producers, who can provide this information. But I also include the politicians and the authorities, who have important roles in ensuring that the information provided by the producers is reliable. My concrete suggestion is that production histories can be communicated through the use of the concept of traceability. Implementing traceability in the food sector could, from the consumers' perspective, be a way of dealing with fragmentation and establishing new coherences and relations.

[312] Taylor, Charles: *The Ethics of Authenticity*, p. 117.
[313] Ibid. p. 77.

Traceability is a concept that has emerged only recently in the food sector and for quite other reasons than I suggested above in the introduction to the chapter. But let me start by introducing the current use of the concept. In general, the introduction of the concept is ascribed to changes in production practices: the separation of production and consumption, the introduction of mass production and the international free exchange of goods. With these changes, new and unforeseen consequences have followed. Errors and shortcomings may appear during the production process, and it is important to be able to trace these, as they may otherwise cause harm not only to a single person, a single family or a small local community but also to a great many people all over the world. Traceability is aimed at being as precise as possible in order to find out exactly what went wrong during the production process so as to make the necessary repairs and withdraw only the malfunctioning products. In some cases it is also essential to find the person responsible in case the faults are caused deliberately. This is the case when health-threatening chemicals are deliberately added to food, or when some participant actors in the chain from farm to fork do not pay enough attention to food hygiene, which can lead to food contaminated with pathogens such as salmonella.

The food sector seems in general to have been less able to see the benefits of implementing traceability and thus slower to introduce the concept than other sectors. However, with the emergence of food scandals such as the outbreak of bovine spongiform encephalopathy (BSE) in the 1980s in England, the discovery of dioxin in animal feed in Belgium in the late 1990s, and the rapid spread of pathogenic organisms such as salmonella, listeria, clostridium and *Escherichia coli* O157 and other contaminants in food, traceability became a major issue in relation to food. The intention is of course to increase food safety.

However, as I have already suggested, there are several other reasons than food safety for using traceability in the food sector. The most common arguments are listed in Table 4.

Food safety and control of production processes are the most widespread aims of traceability. In Europe this is now obligatory, as it is implemented by law in EU regulations. Authentification of product labels and quality assurance schemes is another area where traceability can be very useful. This is mostly done on a voluntary basis in the sense that producers voluntarily adopt such schemes, e.g. organic farming. The last point concerning consumers' perception and interests suggests that traceability could be used to serve specific consumer interests and claims concerning transparency and informed choice and hence also ethical consumption. This is still a highly controversial issue but also at present a hypothetical issue, as no political measures have been taken so far to make food traceability accessible to the general public. The next section shows an example of how the concept of traceability could be used to communicate the production history of foods.

Table 4. Traceability functions

Objectives related to food traceability
1. Food safety and risk assessment • Safety: identification or mapping of breakdowns in food safety allowing recall of contaminated products for the purpose of protecting public safety • Risk assessment: mapping of foods, food ingredients or processing technologies that may have significance with respect to food safety (e.g. hygiene)
2. Control • Surveillance and auditing of producer and retailer activities • Food residue surveillance: food sampling at appropriate points testing for residues, e.g. pesticides • Avoidance of fraud and theft: Control of products by chemical and molecular approaches (biological "foodprints") • Identification of responsible actors (but also claims of innocence!) • Label verification • Ingredients definition • Avoidance of negative claims (e.g. "may contain GMO traces") • Minimization of recall losses
3. Valorization and quality assurance of products • Marketing, health, ethical and other claims • Authenticity: identity of the product (food authentication) and the producer • Quality: final product standard assurance
4. Consumer perception and interests • Transparency • Informed consumer choice • Confidence and loyalty • Recognition of consumer claims and concerns • As a tool for comparison of different products • Public participation: consumer services and companies' "care lines" for consumer feedback

2. SOCIOLOGICAL SURVEY ON ETHICAL TRACEABILITY

In 2004 the concept of traceability was used in a sociological survey into the communication of a production history to customers in six supermarkets in Copenhagen, Denmark.[314] The idea was to analyse how people react to an increase in information on the production history of certain articles available inside the shop. Would this help to bridge the gap between attitude and action as outlined in Chapter One? If customers could trace the ethics of a production

[314] The survey is published in Coff, Christian, Lise Walbom Christiansen and Eva Mikkelsen: *Forbrugere, etik og sporbarhed* (*Consumers, Ethics and Traceability*).

history, would they then act in accordance with their opinion? The expectation was that their choice of articles would be influenced by this information and by ethical traceability. The analysis was not merely to be descriptive, as in showing how customers shop nowadays and how this is or is not related to their opinions; for the gap between attitude and action might well prove to be due to a lack of customer information. The experiment was therefore to be *normative* in the sense that the aim was to analyse *what* was required to make customers act more in accordance with their opinions. In other words, we wished to analyse how to meet in practice the customers' wish to shop ethically. So from focusing mainly on the descriptive, as in a contemporary analysis of the impossibility of political consumption and the strategic application of the concept in the exercise of power, the project became more practical and took on a normative orientation. The question was then: How *could* political consumption become a real possibility?

The setting up of the experiment is described below and is worth noting here, for it shows how hard it is to put together a production history for industrially processed foodstuffs; at the same time it details the information presented to the customers.

First, we had to ascertain the type of information that customers were interested in – and indeed if they were at all interested in knowing anything about traceability and production history. To do this we set up three focus group meetings at which participants exchanged opinions on the ethical significance of production history, on the relevant ethical areas and the kind of information that customers ought to relate to, and finally on what knowledge and tools should be provided to enable customers to act in accordance with their attitudes. Five important criteria resulted:

- There is too much information – we want more information. Many would claim that customers are already overloaded with information. And yet despite the bombardment there are customers who claim that they want more. It sounds like a paradox, but it may well be the case that the information they are receiving is not what is being sought. If the accessible information is not relevant, there is a good reason why customers are both overwhelmed by unimportant time-consuming information, yet feel the need for more.
- I am responsible – the shop is responsible. This too sounds paradoxical, but it reflects the customers' sense of ethical responsibility for their shopping while regarding the shops as being responsible for presenting information on the ethical origins of the products on their shelves.
- Shops are the customers' channel to the producers. The customers' demand for information and ethical action during production is aimed less at the producers than at the shops. Customers regard themselves as pretty much powerless

in relation to the producers, but see certain opportunities for the shops to act on the ethical conditions pertaining to production history.

- A regard for working conditions, animal welfare (including transport) and environmental sustainability are the three areas mentioned most often by participants.
- The need for information on the production history and ethical traceability of the foodstuffs is very individual, making it difficult to reach all consumers with only one type of information. There were many suggestions as to how production history could be mediated, such as: a credible and comprehensive declaration of contents, shop assistants' knowledge of the articles on sale, shop leaflets and posters, use of the Internet, appeal to the senses and feelings instead of text overload; all these being part of a clear ethical policy. All available information should be gathered in one place; the Internet is an obvious possibility. Easy access to this information is a clear customer demand.

With these concerns in mind we compiled information material on the production histories of selected articles and placed it in the six supermarkets in question, using three methods of communication:

(1) Computers with a homepage of pictures and print were installed in every supermarket.
(2) Folders using pictures and print were readily available.
(3) Demonstrators were present who were versed in the production history of the articles and able to inform customers about them.

This means that there were two shops with computers, two with folders and two with demonstrators. The two products used in the survey were rye bread and liver pâté – two Danish staples with a large turnover in supermarkets.

The biggest task for the setting up of the survey unexpectedly turned out to be the communication of traceability and production history. First, it was important that the production histories contained information on brands within the same article so that they could be fairly compared. It proved to be harder to standardize the production histories than was assumed. Second, information had to be extracted from the producers, which was also more complicated than we had imagined.

The standardization of production histories is important for comparison of the products and for offering a fair choice. We had a list from the focus groups with subjects that they considered relevant, providing a guideline to the content of traceability and production history from an ethical point of view:

(1) **Geographical origin**
(2) **Product quality**: Description of manufacturing methods and their effect on the quality of the product (e.g. post-harvest treatments and processing techniques)

(3) **Production and environment**: Description of environmental initiatives and possible certifications of the product
(4) **Production and animal welfare**: How are animals treated?
(5) **Production and working conditions**: Has anything been done in particular to ensure good working conditions?
(6) **Production and society**: Description of farms and businesses that are involved
(7) **Economic transparency**: Who earns what? Economic fairness as an ethical parameter
(8) **Gaurantees and certifications**: Who is responsible for ensuring that the information on the article is correct? Customers must have confidence in the relevant inspection authorities.

On the basis of this list we drew up a standard questionnaire, which we asked the producers to fill out – a complicated matter, it transpired: some were unwilling to provide us with required information; one of these, Danish Crown, has since been the subject of media attention for allegedly employing underpaid Polish workers in its German factories.

It also appeared that some of the actual questions were problematic. We had to drop completely the one on economic transparency and fairness, since there were far too many factors at work. For example, in the case of liver pâté, the question of who earns what is highly relevant, since many customers suspect the major producers of squeezing out the small ones. It is easy to calculate the supermarkets and the producers' share, but what about the farmer? Farmers do not sell liver; they sell whole animals from which the liver is extracted at the slaughterhouses and sold separately to liver pâté producers. The slaughterhouses pay the farmers for whole animals and not separate parts, so nobody could reasonably calculate what the farmer earns for the liver. Moreover, incomes must be seen in relation to the expenses of the individual parts. Large incomes in themselves are not at issue when accompanied by large expenses. Economic fairness must thus be developed in quite a different way if we are to make sense of production history.

For those businesses that wished to join the survey there were major differences in the information they possessed. In general, producers of organic products provided the most information, which is not surprising since they are subject to a voluntary market arrangement that demands a certain degree of documentation. But a number of businesses found it impossible to answer the questions; they could only state that current legislation on animal welfare, the environment and working conditions was being observed. These empirical findings indeed confirm the fragmentation of the food market. Companies possessed generally very little information on the history of the products outside their own company. They also underline the "hiddenness" of the production history as an empirical fact. So we had a problem: the letter of the law is hardly the best message to mediate. It seemed that the businesses barely had a story to tell about their products, as these were so anonymous. And in the case of the most industrialized products, we could learn nothing about the origin or growth of the animals, about working conditions at the farms, about manufacturers or about their environmental concerns.

A third problem turned out to be the widespread market dominance by a few businesses. We had three different brands of rye bread in our survey, all of which turned out to be based on flour from the same miller, Cerealia Mills. The market for flour in Denmark is dominated by three large businesses, Cerealia being the biggest. This presented us with two problems. First, the flour could not be traced further back than to the mill where all the grains were mixed – from seven different European countries. Nor could we trace the farm in question, the region or the country of origin. This was a pity, since geographical origin is important for many people, who link a given foodstuff to a particular geographical region, landscape or culture. Second, this made the production histories very similar. What appeared to customers to be three different brands of rye bread were in fact very much alike. Indeed, in one of the supermarket chains the 10–12 different brands all stemmed from the same bread producer and thereby defeated our purpose: if there is no difference in production history, only price and taste remain as delineators, and ethics are irrelevant. Since businesses are not geared to distinguishing and recording production histories, they cannot provide relevant information. In consequence, our own material was less detailed than desired. Figure 2 shows an example of how the production history of bread was mediated via the Internet. In two of the six supermarkets, computers were available for customer use.

The production histories presented were brief in order not to burden customers more than necessary. Below is the history of Hedebagerens Rugbrød (Heathbaker's Rye Bread) from the above Internet page:

The Farm

The rye is cultivated by ecological growers in Denmark and abroad. This means that no synthetic pesticides or fertilizers have been used, only mechanical weeding, seed changes to avoid plant diseases and animal and "green" manure.

The farmers deliver directly to the miller (Cerealia Mills), who can trace the grain to the individual farm. However, once the ryes are mixed and ground they can no longer be traced back to the individual farms, so there is no information about working conditions on the farms. The farmers sign contracts with Cerealia Mills for delivery of the rye, which is traded at market price.

The Mill

The rye is ground at Drabæksmølle (Drabaeks Mill) in Lunderskov in south-central Jutland. The rye is a 100% ground full-grain product.

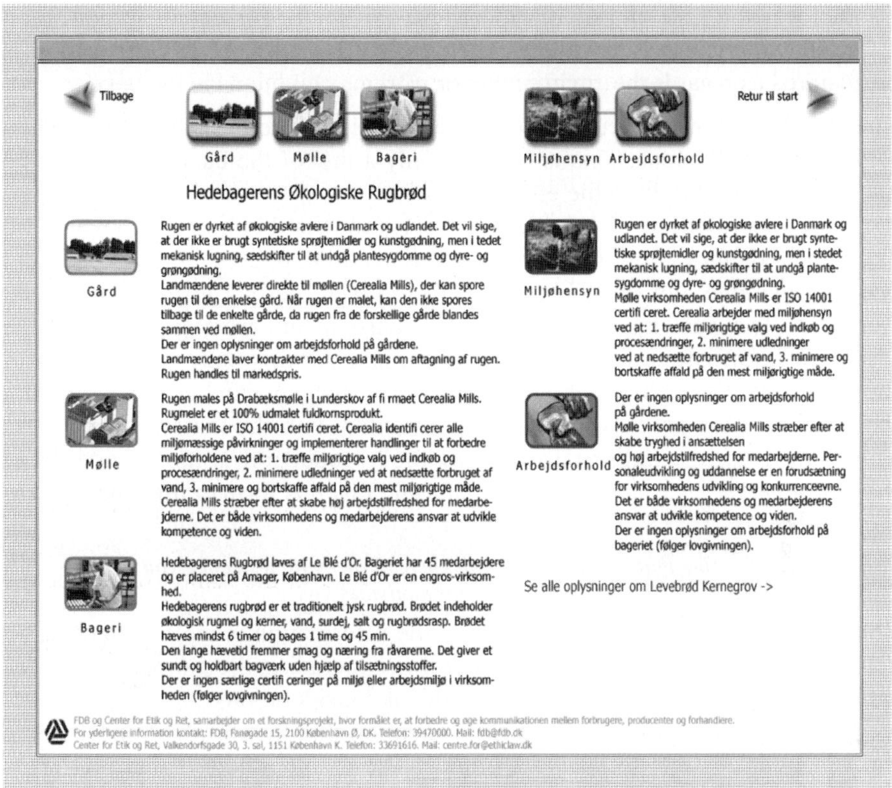

Figure 2. Example in Danish of the Internet communication of the production history of the bread. (See accompanying English text below.)

Cerealia Mills is ISO 14001 certified. It identifies all environmental effects and implements measures to improve conditions by: (1) making correct environmental choices and processing changes; (2) minimizing discharge by reducing the use of water; (3) minimizing and removing waste in the most environment-friendly way. Cerealia Mills aims to create a high level of job satisfaction among its staff. It is the responsibility of both the business and the staff to develop skills and knowledge.

The Bakery

Hedebagerens Rye Bread is made by the bakery Le Blé d'Or (The Corn of Gold) in Amager, Copenhagen. This is a wholesale concern.

The bread is a traditional Jutland rye bread containing organic rye flour and kernels, water, leaven (sourdough), salt and rye bread crumbs. The bread rises for at least 6 hours and is baked for 1 hour and 45 minutes. The long bread-rise

> improves the taste and nourishment levels of the raw materials, producing a healthy, durable bread without additives. By law no *special* certification is required as regards the environment or working conditions in the bakery.

The text shows how difficult it is to place a specifically ethical angle on the history of the productions. This is because only a few of the businesses involved had an explicit attitude to the ethical question.

In the supermarkets we carried out interviews with the shoppers in which we asked them their opinion of the information material. Interest in the experiment was considerable: out of 257 questioned, 155 were willing to participate, 102 were not. The vast majority of the 155 participants were grateful for the increased information despite the shortcomings and the many similarities, for they knew next to nothing about production history. A number of them had a vague idea of the process, but when confronted with the material, they were often surprised; the industrialization of food production is more widespread than most people imagine. One informant said that this form of information was *motivating*:

> *I think it's fine that the information is there. Especially if like me you're not very good at finding out about it – so it motivates you.*

The customer does not add in what way the information is motivating, but we can imagine that there are several possibilities including the motivation to *seek* information and to *act* on the basis of ethical considerations.

Many customers were also quick to mention that the information available must make it possible to *compare* brands. The problem here is that producers are unwilling to produce information that may be regarded as negative by customers. So politicians, businesses and authorities also have a role to play, for it cannot just be left to the free market to inform customers about the production history of the articles, since all they will get is the usual PR material.

The survey also showed that ethics are clearly regarded as a difficult and diffuse concept. This may be due among other things to globalization. For modern ethics is not just about the short range and what we immediately understand. As described in Chapter Four it forces us to relate to the long-range consequences of our actions that take place on the other side of the globe. It is precisely here that the difficulty lies, for we do not know these consequences.

The interviews carried out in the survey point to a number of concrete conditions that are of importance for the communication of the ethical traceability of the foodstuffs. The main results are analysed in Table 5.

The main conclusions of the survey are, first, that increased information on article traceability arouses considerable customer interest. Many respondents think they need help to find their way through the consumer jungle. Second, the introduction of the practice of ethical traceability and information on the production history of the articles cannot be expected to show immediate results among customers. Our unsurprising experience was that it takes a long time for

Table 5. Main results from an empirical survey on ethical traceability

---

INFORMATION

- Information on production history is clearly **new information** for most respondents.
- Customers will very rarely be able to find information on ethical traceability but **require help and guidance**.
- It is often mentioned that there is overinformation on the foodstuffs market. This survey points rather to **lack of information** as an equally possible problem: that the available information is not what is being requested.
- **A branding system is used for swift information** from day to day, but information as to what it stands for is not readily available; more information is required on the different brands.
- Information on foodstuffs and their ethical traceability must be **easily available** and not consist of lengthy, heavy texts. It should also be provided **at the point of sale**.
- For many customers **sensuality** towards the foodstuffs plays a role, but they are often excluded from this opportunity in the supermarkets.

COMPARABILITY

- It is crucial for customers to be able to **compare brands**. Differences must be highlighted to facilitate customers' choices.
- When the only comparable parameter available is price, it is natural that **price becomes the dominant factor** for many customers.
- **Cheap offers** make the market less transparent, since customers lose the sense of what various foodstuffs normally cost.

ETHICAL TRACEABILITY

- Traceability of foodstuffs is chiefly linked to the **primary place of production**, i.e. the soil and landscape from which the food derives. For many customers the farm is more important than the processing factory.
- **Geographical indication** is not merely the specification of *place*, but may have political, ethical, cultural, social or environmental significance.

TRUST

- A number of customers have **great trust in the retail chain**. This can be explained by their lack of contact to the producers.
- **Transparency**, in the form of publicly available information and criteria for the presentation of information, creates trust.
- **Impartiality** among information providers is regarded as essential by practically every respondent.

ETHICS

- Just over half of those interviewed thought that **increased information on the production history and ethics behind the foodstuffs would change their shopping practices.**
- Information on traceability and ethics has a **motivating effect** in relation to shopping with an attitude.
- The customers' interest in the ethics of production history concerns in particular **animal welfare, the environment, working conditions and fair trade**.
- There are great expectations that **shops** will take ethical responsibility for the merit of their articles and for information to their customers.

---

customers to get used to such information being available and even longer to apply it to their shopping practices. Since consumption is so much a matter of choosing, it can be regarded as a learning process.[315] Consumption is not automatic; it is based on experiences, exchange of opinions, surveys and so on, not to mention advertising. Because ethics since the arrival of modern consumption has been almost excluded from the market, as we shall see, and in consequence removed to the political sphere (at the very least theoretically), it will require a major revolution to make the ethical factor an integral part of the market. So the opposition it encounters is not surprising. Customers are used to ethical topics on production practices turning up in the media, but to be presented with them directly in the shop is an innovation. An example of this is presented below.

Woman in her thirties with a child:

Q: *[demonstrating use of computer to customer] Do you think it's something you'll use?*

A: *No, I won't. The only thing I'm interested in is what there is in the way of additives. I'm only interested in the content of the articles actually . . . . So here and now I'd say no – not if it's things like this that are in the database.*

Q: *Do you think there's enough information on the articles as it is today?*

A: *Oh, [thinks] if I say that what I need to know is what kind of contents are in it, then I don't always get the information I want.*

Q: *But doesn't it always say what's in it?*

A: *No, but er, let's see here [looks for something on the computer] – it doesn't say, for example, whether they've used growth inhibitors.*

Here we can see how at first the respondent shows little interest in the production history, but nevertheless is concerned about the use of chemicals in agriculture since they reappear in the foodstuffs. Perhaps she has been surprised to find a link between the content of the food and the way it has been produced.

The conversation also reveals *who* the consumer perceives as owning the information in the foodstuffs market, and who therefore is able to set the agenda as to what is, and how it is, passed on. Information regarding production history, ethics and traceability are new subjects that only a critical and committed minority of consumers can be expected to search out and take an explicit stand on. Most consumers would not immediately be able to relate to a production history until they were presented with one. In the first instance, it is not so much the consumers as the producers and dealers who would be able to initiate such a development. But for this to last and to stabilize it must of necessity enjoy consumer backing.

[315] Sassatelli, Roberta: "Tamed Hedonism: Choice, Desires and Deviant Pleasures".

Traceability proved in the survey to be a concept with more nuances than at first assumed. Geographical indication, which in itself is no more than a simple proper name, often acquires far more significance than it immediately indicates. For by virtue of the individual images customers have of specific lands and regions, many of those interviewed are influenced by the working conditions, animal welfare and regard for nature in general. We now jump into the middle of one of the interviews:

Woman in her fifties:
Q: *Things like vegetables, it doesn't often say where they're from.*
A: *Well it should do.*
Q: *Is that something you think is important?*
A: *Very. Extremely important. For example, if I want to boycott articles from Israel, I sometimes have a few problems there, because there are false labels on them. Sometimes they write that they're from South Africa when in fact they're from Israel. But I look at the boxes to see where they come from!*
Q: *But would you use it [traceability] to do the same with new articles?*
A: *Yes, I think it's a good idea, if for no other reason then to draw greater attention to it.*

Geographical indication is not just a matter of *place*; the place of origin is linked to a number of other factors mentioned in Table 5: political, ethical, cultural, social, environmental and so on. The boycott of Israel is an instance, but it also became increasingly clear in the course of the interviews that working conditions were not regarded as a problem for Danish products, for it is believed that Danish legislation has laid down firm regulations. This indicates that geographical indication is "overinterpreted" in the sense that consumers tend to read more into the information than they are actually given. "Produced in Denmark" refers to a country with strict regulations as regards working conditions, but in reality working conditions need not be that ideal. The speed of work, for instance, is often mentioned as a stress factor for Danish workers.

The survey also shows that the primary place of production is far more important than the secondary place of processing. This can be interpreted to mean foodstuffs and traceability are first and foremost linked to the nature and the "landscape" that produces the food.

3. CONSUMER AUTONOMY: REMEMBERING THE OTHER AND INFORMED CHOICE

The interest in ethical consumption, which is growing despite the wave of corner-cutting discount articles, is beginning to cause a political stir in an otherwise unlikely area. The dominant liberal understanding of the market is that it is private, not political – regard for oneself takes precedence over regard for others. But ethical consumers seem to be ignoring this, leading us to question how the liberal view that separates markets from politics so fundamentally has managed to maintain its dominance.

In pursuit of a similar agenda the Dutch philosopher Michiel Korthals explains convincingly in his book *Before Dinner* how the distinction between consumer and citizen arose in the Age of Enlightenment as a rejection of feudalism. In feudal society, production and trade were not free; they were controlled by political interests. Princes did not distinguish between their own private finances and the economic interest of the principality or dukedom they controlled: the two were identical. Production and trade were subject to political frameworks and conditions.

The emergence of the concept of citizen in the 18th century is very much part of the rejection of that social structure and finds expression in the idea of *civil rights*, which are intended to protect citizens from the injustice perpetrated on the individual by feudalism. Civil rights confer on citizens a sphere that is both private and protected from state interference. Under the influence of Rousseau's distinction between "citoyen" and "bourgeois", incipient liberalism develops a view in which consumers (the bourgeois)

> are assigned a personal public and private domain, where their actions have consequences only for themselves; the market . . . . Citizens (*citoyens*) come together and act on the basis of civil sovereignty.[316]

Due to the historical link in the liberal tradition between consumer culture and the protected individual and private sphere, consumption is often associated with the liberal notion of choice, often articulated as consumers' *free choice*. Because the market is thought of as being demand-driven, the free choice of consumers has become a central concept of liberal economic theories as to how the market functions. But what is actually meant by "free choice"? This is no longer apparent, since most consumers in the industrialized countries already feel themselves to be free; or at the least not directly constrained. The opposite of free choice would be "forced choice" – which in truth is not a choice but a forced action, a kind of consumption where one does not choose oneself, but where others or "necessity" makes the choice for one. The latter was, for instance, the case in earlier times when the tyranny of the seasons subjected people to a very local food regime. Today, forced choice would rather imply that others decide what kind of products one should buy; one would, for instance, be forced to buy and consume a specific kind of milk. Of course this seems absurd. In a less radical and more reasonable sense, a "forced choice" would mean that one would have no alternative

[316] Korthals, Michiel: *Before Dinner*, p. 155.

between a specific brand of milk and no milk at all. When there is a monopoly, one is forced to buy milk from that specific company or not to have any milk at all (or make it oneself, which is an option only for the few). In a more subtle way, "forced choice" could also be an indication that consumers are manipulated, e.g. by advertising, or are subject to social pressures to buy certain products.

These reflections on forced choice and the role of advertising lead us more deeply into the concept of freedom and what we should understand by this. Advertising can be regarded in two opposing ways: manipulation and information. As manipulation it hinders free choice, whereas as information it helps to make the choice more free. But in what way can such a choice be "free" at all?

Freedom is probably the most important principle in the tradition of civil rights. Civil rights have a long history dating back to the *jus civile*, the law of Roman citizens. The English term "civil right" was coined in the late 15th century and was first used to distinguish civil law from canon law. Well-known examples of civil rights are the French *Declaration of the Rights of Man* (1789) and the *Universal Declaration of Human Rights* (United Nations, 1948). The list of rights has grown since the first declarations and today it covers many areas: the right to self-determination, the right to life, the right to liberty and freedom of movement, the right to equality before the law, the right to presumption of innocence until proven guilty, the right to privacy and protection of that privacy by law, freedom of thought, conscience, and religion, freedom of opinion and expression, freedom of assembly and association and so on. Discrimination against an individual occurs when these rights are denied or interfered with. Rights offer a safeguard against encroachments like violence and injustice.

Not only has the number of rights (or freedoms) increased since the first declarations, so has the number of citizens *possessing* those rights. Today, not only free men but also women, children, the disabled, patients, animals and so forth have their own rights. Some would even speak of the rights of ecosystems. In jurisprudence it has become common practice to speak of three generations of rights: the liberal civil rights of the 18th century, which mostly describe the negative rights securing the liberty of citizens; political rights of the 19th century describing the rights to participation in politics; and the social rights of the 20th century on the just distribution of goods in society and equal rights to welfare. Today, a fourth generation of rights is being developed. These rights typically concern issues like health and environment. Also, and this is not surprising for "the information society", communicative rights or the right to know is being addressed as central for recognizing the autonomy of individuals and for maintaining their rights to participation in political processes in society. The rights of ethical consumers are to be found in this not yet fully developed fourth generation of rights. Protecting consumer rights is an issue of growing importance, both at national and international levels. Linked to the abovementioned traditional liberal distinction between market and politics, the formulation of rights in trade relations is another evident sign of the invasion of politics into trade and vice versa, another clear indication that the implications of trade cannot be separated from the political sphere and that trade has implications for civil society as well.

However, the concept of individual freedom or liberty is not only linked to the liberal tradition that developed in opposition to the feudal regimes of the Middle Ages in Europe. To gain a fuller understanding of the concept of freedom it is useful to consider the term more broadly in the light of the philosophical tradition. The concept of individual freedom is based on the normative idea of individual autonomy. As a philosophical concept autonomy does not designate a single thing: there are various philosophical understandings of this complex concept. Autonomy (Greek: *auto–nomos*) literally means self-rule, self-determination or self-government and in this sense it was originally used normatively in the description of sovereign (autonomous) nations. With Kant the concept was given a strong twist towards the individual. The idea of individual autonomy is normative in the sense that it is not something naturally given, but is rather used to describe a vision of an ideal person and how that person makes decisions. Autonomy is not something inherent in man, but rather something to strive for and which is facilitated through education and cultivation (German: *Bildung*).

As part of the vision of the good life the normative concept of autonomy is based on the negative experience of personal infringement or injustice when being forced or determined by external conditions against one's own will. Individual autonomy as one of the oldest and most fundamental principles of civil rights is intended to protect against infringements; it is the right of the individual to be self-legislative and to make decisions without coercion. Therefore, at a very general level, an autonomous action is a voluntary action that is not forced by external conditions. Hence, *independence* is traditionally considered a core value of autonomy. Autonomy is the capacity for independent decisions, choices and actions – the independence from others, or from others' views or preferences.[317] This is often referred to as negative freedom: the freedom from interference by others or the government.

To clarify the concept of autonomy the American philosopher Thomas Hill distinguishes between three different layers or kinds of individual autonomy that can be found in the history of philosophy:

[317] O'Neill, Onora: *Autonomy and Trust in Bioethics*, pp. 23–24.

- Moral philosophy
- Jurisprudence
- Philosophical theory of the person[318]

Let me start with the second layer, jurisprudence. The negative-freedom interpretation of autonomy belongs according to Hill to the legal sphere, since it describes the legal rights of citizens. The independence of individuals is of course within limits, as the autonomy of other persons must be respected likewise. In a well-organized society there must necessarily be restrictions and regulations on individual autonomy. The most obvious of these is probably the principle of non-injury. Nor are human subjects free to cheat each other or to disrespect the private property of other persons. And yet the legal right to autonomy does not say much about the actual act of self-rule or the individual capacity for self-government, as it only protects the rights of the individual.

This leads us to the first layer of moral philosophy, in which there are two different understandings of individual autonomy, both relating to decision-making and judgement. The first understanding is from Kant, in which individual autonomy is the capacity of a person to be his own moral legislator, free from sensuous desires, traditions, personal inclinations, etc., in order to make moral judgements based on rational principles. Justification and argumentation is central to the moral judgement. The second understanding of individual autonomy emphasizes impartiality. According to this understanding, justification by argument leads to the considering of principles in practice and is thus a detachment from personal inclinations. A moral judgement is distinguished from other judgements and is said to be autonomous when it is based on rational argumentation and impartiality. Morality is thus not necessarily a characteristic of action, which can be unreflected and spontaneous; more specifically, it is to be found in the rational and impartial procedures of judgement.

The third layer of individual autonomy concerns the self-awareness of the moral agents and the conception of the human subject. Self-governing individuals cannot be blind to pressing problems of a given situation, rational arguments, the value of values, their own needs, etc., because that would be a contradiction in terms. Instead they must be able to acknowledge their own values and intentions: the human subject must be transparent to himself.

The question now, and in the rest of the book, is how these concepts of autonomy apply to the choice of the consumers and to the concept of ethical consumption. As we have seen, the exercise of choice clearly requires the ability

[318] Hill, Thomas E.: *Autonomy and Self-respect*, pp. 43–51. See also Honneth, Axel: "Decentered Autonomy. The Subject after the Fall" for Honneth's interpretation of Hill's distinctions. In Honneth, Axel: *The Fragmented World of the Social*, pp. 261–271.

to reason, reflect and judge. Otherwise it would hardly be considered a free action but rather the result of instinctive behaviour.

Let me start this inquiry by considering the question of whether choices can be completely free and without commitments. First of all, it is doubtful if the unconditional free choice, where the human being is totally unrestrained, is possible. Completely free choice is only imaginable at a theoretical level; at a practical level it is hardly possible. Apart from disrespecting the autonomy of others, the unrestrained free choice has also been criticized by many philosophers, including Hegel, who thought that the outcome of this kind of freedom would lead to a "suffering by indefiniteness". Free choice in its most radical form is undesirable, for it neglects the interests of others and of society as a whole. Free choice is thus a choice made in isolation that also enforces personal isolation. In spite of this, free choice, in the pessimistic view of Charles Taylor, is a major characteristic of Western societies because "its spirit is an adversarial one in which citizen efficacy consists in being able to get your right, whatever the consequences for the whole".[319] Choosing freely has become a value in itself, no matter what the choice is; the consequences of the choices are of far less importance. This interpretation of choice, centred on the maximizing of utility and on the egocentric self, has become a common way of understanding the term. For this reason the concept of choice, especially when considered in the context of consumption, is often treated in a rather condescending manner. Choosing relates to the satisfaction of one's own desires where the needs of others do not come to one's attention.

However, choosing, even in the context of consumption, is not necessarily such a simple activity. Choosing might just as well be highly reflective, based on careful judgements and evaluated arguments. Choosing can be a relational activity that takes consequences into consideration and connects people with their surroundings. The question addressed here, then, is whether the concept of consumer free choice, where the focus is on free choice in the liberal sense of self-governance and independence, properly addresses consumer needs.

My claim is that most consumers already do think of themselves as free in the sense of *being able to make unforced choices*; in spite of the wealth of advertising nobody directly *forces* the consumers to buy specific products. Modern consumers are indeed free to choose among available commodities – or not to choose certain products. There are financial limits to people's purchasing power, but in most cases no one is forced to buy specific commodities if they really do not want to. Only in situations of monopoly, when, for instance, there is only one company providing milk, can we say that the consumers have no free choice. In this case *free choice* can serve as a means of protecting consumers against monopoly. Thus, the liberal demand on freedom of choice seems

[319] Taylor, Charles: *The Ethics of Authenticity*, p. 117.

in many respects to be anticipated already. It is nevertheless a concept and practice that we should continue to defend and safeguard.

However, even though consumers are indeed unrestrained in this sense of freedom of choice, they are certainly not free in another sense; for the freedom to access information on food and production practices is certainly not respected. The sociological investigation referred to above shows how difficult it is to obtain information on the production history of food. And even though freedom is a major substantive civil right, the procedural right to free access of information does not exist. The autonomy of consumers should not only be understood as a substantive right, i.e. as the right to simple free choice, but also as a procedural right for ethical consumers with regard to the access to impartial and reliable information on the production history.

In today's complex society, information and knowledge are key elements for making free choices. I would argue that choosing without knowledge is not a free choice. It is only free from the perspective of the egoistic utility-maximizing consumer, but from the perspective of consumers who see consumption as a way of relating to others it is a coerced choice because it is not known what is chosen. Thus, the narrow understanding of autonomy as *free choice* is inadequate and should in the context of ethical consumption be interpreted in another way. For the sake of the analogy it is worthwhile dwelling for a moment on the interpretation used in medical ethics, e.g. in the Helsinki Declarations (1948, revised in 1975) on Human Rights. Respect for patient autonomy protects the right of the patient to choose the medical treatment, which he or she prefers. This must of course be done on the basis of information on the illness and the possible treatments provided by a physician. Autonomy in medical ethics is commonly practised as freely given *informed consent*. This example underlines, without going into all the practical problems involved in the use of informed consent in medical practice, the importance of information and knowledge in making free choices.

Within consumption we do not speak of informed consent but of *informed choices*, which indicates that the choices are based on information. The coercion of consumers is brought about by the lack of information or, as is even worse but all too frequent, misleading information. An indication that this is actually what is taking place is seen in the belief by most consumers that their attempts to practise ethical consumption are an illusion – due to the lack of information and/or misleading information.

For ethical consumers, autonomy as independence or unrestrained choice is of course vital. Yet there is more to it than that: ethical consumption is a *relational* or *connecting* activity based on the respect and care for others. This is especially evident in the area of food. Cooking and sharing a meal are fundamental ways of showing care, typically within the family or among friends. This is also why providing food that is rejected is disheartening and disappointing. Even though the British sociologist Alan Warde argues as I do that food is increasingly being regarded as a technical matter, he also stresses that cooking is strongly linked to the care for others.[320] Choosing food is a relational activity in that the consequences of the actual choice are taken into consideration: choosing connects people to their surroundings, not only to relatives but also in a broader sense to society and nature. To illustrate the discussion on autonomy and the understanding of the concept among consumers, five statements from the traceability survey mentioned in Section 2 are presented below:

- Consumers must make claims. We are free to choose whatever commodities we prefer.
- We should take a stand on all that is possible. Our attitude should be based on the available information.
- Consumers can make a difference via their purchasing power. We decide where and on what to spend their money.
- It is a problem that there is no real choice between brands. On the face of it there are many different brands, but in reality only a few companies are responsible for the products.
- All those involved in the production process must be treated in the best possible way. All links in the chain – human beings, animals and plants – must be safeguarded (a tall order!).

These statements indicate several aspects of food consumption. First, consumers actually feel free to choose what they want. Free choice is more or less taken for granted by consumers in the survey. Second, the importance of information for making choices is stressed. In spite of their free choice, shopping for food is experienced as far from problem-free. Even though choices are free, consumers do not experience themselves as self-determining. The lack of information is too vast to speak of an actual self-determination of consumption. As regards this right, it is outmoded merely to refer to the free market and traditional liberal freedom of choice. When freedom of choice is the limbo that dominates the market, there is a strong risk of resignation and the abandonment of political and ethical consumption. Third, these statements highlight consumption as an ethical activity because ethical consumers consider the well-being of others, including nature, as part of their shopping. They link their consumption to a responsibility towards society and nature because it mediates a relationship to them, which they believe all consumers should admit. Ethical consumers feel

[320] Warde, Alan: *Consumption, Food and Taste*, p. 130. Warde is right in that cooking has also been used in a subtle way to subordinate women by repressing their own desires and making them deferential in the pleasing of others.

that a kind of responsibility towards nature and society should be admitted in relation to their consumption.

The fact that consumption has implications for society and nature makes consumption political and ethical; whether we acknowledge this or not, consumption still has consequences that are important in political and ethical perspectives. Thus, consumption is both an economic activity and an ethical and political activity. Markets and politics are indissolubly linked. People are beginning to regard themselves both as consumers and as citizens and their consumption to be not only a private but also as a public matter. The traditional liberal distinction between consumer and citizen is outdated because buying behaviour and political interests no longer belong to separate spheres. Individuals act in both the political sphere and on the market. The widespread academic distinction between citizen and consumer was found in only few of our respondents in the survey; the clear interest for greater knowledge of the production history is proof of the blending of politics or ethics and the market, and of the citizen and the consumer.

4. RECOGNIZING CONSUMERS RECOGNIZING PRODUCERS

In Section 3, I wrote about the need for viewing the autonomy of consumers also as a procedural right, i.e. as a right to impartial information on the production history as the basis for consumers' informed choice. In democratic societies there is more to autonomy than self-determination and rights: autonomy also presupposes recognition, without which it is worthless. Or, the other way round, to recognize somebody is often taken to mean to respect the autonomy of that person. Thus, in a liberal and democratic society characterized by constant interaction between people, autonomy is closely linked to the concept of recognition. In practice, autonomy without recognition from others is impossible because without recognition one is likely to be ignored, forced or manipulated by others or external conditions.

To make the point clear let us consider a case where autonomy is not linked to recognition: the case of tyranny. A tyrant or dictator does not need his autonomy to be recognized, as he does whatever he likes, whatever others may think or do about it. He is absolutely self-legislating and independent of others and as such has no need to care about or respect others. The leadership of the tyrant is based not on recognition of his leadership but rather on repression and the subjects' fear. In the case of the tyrant, autonomy is one-sided, as it is only ascribed to one person while all others are excluded.

However, in democratic and pluralistic societies based on equal rights and on the autonomy of all citizens, the exercise of individual autonomy requires not only that the individuals possess the capacity to behave autonomously but also that the autonomy of others is likewise recognized. In a societal context this questions the rationality of individual autonomy, as it seems to carry an inner contradiction: how to maintain personal freedom without interfering or limiting the freedom of others? This seems almost an inherent conflict in the societal concept of autonomy. On the one hand, individual decisions and actions should be unrestrained by others, yet on the other hand, there is a duty not to interfere with others in order to respect their autonomy. The latter would mean that autonomy is restricted when other people are harmed or limited in their free choice. This is a serious limitation that seems to contradict the idea of self-legislation.

Theories of recognition are often used to address this fundamental and intrinsic problem of autonomy. Autonomy and recognition represent the two sides of the conflict: the freedom of the individual and the freedom of others. In his critical analysis *Against Autonomy* the British philosopher Neal Curtis writes:

> For many theorists it is the concept of recognition that provides a solution to this difficulty [individual freedom versus the freedom of others], arguing as it does for autonomy conceived intersubjectively through mutual relations with others. Recognition is thus the individual conceived socially.[321]

I shall briefly describe some of the other recent critiques against autonomy, as this is important for the understanding of consumer autonomy.[322] Since the 1960s the concept and project of autonomy has been criticized among others from the view of psychology and linguistic philosophy. This means that although autonomy is basic to democratic society and universal human rights, it cannot be considered an entirely straightforward concept. The claim for transparency of the self has been thoroughly criticized; human beings are not transparent and are not always aware of their own intentions due to their unconscious psychological motives. The rationality, reason and reflexivity usually ascribed to the subject are called into question as a result of unconscious and libidinal forces of the mind. The mind is thought of as non-transparent to itself. The will becomes more or less dominated by uncontrollable forces such as unconscious psychological motives and thus makes of autonomy an illusion. Also subjectivity, in the sense of self-determination, was questioned and deconstructed by linguistic philosophy inasmuch as linguistic meanings and systems were thought to precede and overrule individual intentionality. When self-determination is dissolved, autonomy becomes meaningless. Powers beyond the subject's own control were ascribed such overwhelming influence that the content of "autonomy"

321 Curtis, Neal: *Against Autonomy*, p. 95. The theorists and works Curtis refers to are Alexander Kojève: *Introduction to the Reading of Hegel*; Axel Honneth: *The Struggle for Recognition*; Robert Williams: *Hegel's Ethics of Recognition*; Francis Fukuyama: *The End of History and The Last Man*.

322 This critique is based on Honneth, Axel: "Decentered Autonomy: The Subject after the Fall". In *The Fragmented World of the Social*, pp. 261–272.

was considered empty. For ethical consumption this would imply that even in cases where consumers do argue rationally for their moral judgement of a production history, unconscious forces might be at play and "distort" the judgement, so that it would be false to speak of an "autonomous" judgement. For the ethical consumer it would be important to limit such distortions but, as the theories say, this is precisely impossible. I do believe, however, that it would be a serious overinterpretation of the critical theories mentioned to deny completely the existence of rational and autonomous judgements. In spite of the theories there is still a sphere in which it makes sense to speak of autonomy and rationality – otherwise a whole host of things would seem worthless. Rationality can be considered a relational term, describing the relevance of arguments and choices to reach a specified aim (no matter whether the aim is in itself rational or not). Rationality is in this way not abstract principles and rules for argumentation detached from everything, but is linked to the achievement of specific goals. And I do find it hard to believe that all goals are decided upon by unconscious forces: some goals can also be argued for in public with regard to visions of the good life and there might even be agreement on certain issues. In my view there are many reasons not to abandon the concept of autonomy, as it does describe a fundamental value of society; there might however be good reasons for a new interpretation of the concept that pays more attention to intersubjective relations. As is clear from Section 3, I argue in the tradition of pluralism for an interpretation of moral autonomy that takes other persons' equal rights into consideration as well.

The relationship between consumers and producers or retailers is in theory a mutual and intersubjective one. However, in practice I would prefer to say that it *could be* such an intersubjective relation based on mutual recognition, because, as shown in the previous chapters, the contact between consumers and other actors on the market is at present far too restricted to actually include anything like recognition, or even cognition. I shall therefore speak of this relationship in a more visionary or imaginative sense, with regard to how it should be developed in order to take ethical consumption seriously. First of all, this relationship cannot be based on a one-sided recognition of autonomy, as in the case of the tyrant, but has to build on mutual recognition. It should be possible to recognize both (1) producers and retailers and (2) consumers.

(1) There can be no doubt that ethical consumers feel a need to approve of the ways food is produced. Ethical consumers believe that production histories of food should be "cognitized" by consumers, i.e. the production history should be transparent and accessible to consumers, so that they can either recognize or disapprove of foodstuffs depending on the production practices used. Recognition or disapproval should be based on an evaluation of the way the food is produced. Recognizing the production history of the food

from a consumer's perspective implies therefore also recognizing the producers and retailers and their practices of producing and handling food. The question to be addressed here is what *kind* of recognition is involved when consumers recognize the production history and the producers and retailers of an article.

(2) Recognition of the individual autonomy of consumers must be the first criterion to be fulfilled if there is to be anything like ethical consumption at all. Without recognition of the consumers' autonomy they are unable to choose according to their own moral concerns and convictions. The question addressed here is in what way the politics and philosophy of recognition (to which I shall return) would be able to contribute to the clarification of the problematic concept of consumer autonomy and ethical consumption: in other words, how can the individual autonomy of ethical consumers be recognized? It also addresses the question of by whom does the individual autonomy of consumers need to be recognized?

So far we have sketched the two questions related to an intersubjective and mutual recognition between consumers on the one side and producers and retailers on the other. The answer that I shall try to give in the following is an investigation of how the social philosophy of recognition can be used for the analysis of the present market situation and the possibilities for developing new relations between producers, retailers and consumers based on recognition. In doing so I shall draw mainly on the theoretical writings on recognition by the German philosopher Axel Honneth (b. 1949), starting with a short introduction to his social philosophy of recognition.

According to Honneth, it is in the writings of the German philosopher Georg Wilhelm Friedrich Hegel (1770–1831) that we find some of the earliest and most inspiring historical sources and philosophical treatments of the concept of recognition, most prominently in his early writings like *System der Sittlichkeit* and *Jenaer Realphilosophie*. Hegel's philosophy of recognition is usually seen as a reinterpretation of the thoughts of natural rights philosophers of the 16th century like Hobbes and Machiavelli, who saw society as dominated by conflicts and fights, the struggle of man against man. Instead of seeing society as a primitive social struggle among individuals, it was Hegel's idea that in the process of socializing – in children's relations to parents and later on to others – individuals experience other human beings not only as struggling but also (or even more so) as loving and caring. From this experience Hegel makes "the struggle for recognition" a central concept in his philosophy of society and intersubjectivity. It is a main thesis of this theory that it is through recognizing and being recognized by others that individuals become aware of themselves. The encounter with other individuals and their opinions of oneself make it possible to "displace" oneself, to put oneself in the place of the other. According to Hegel, viewing oneself through the eyes of the other is central to self-understanding and thus also to identity formation. Self-awareness is formed in the presence of the other. It is also by being recognized by others that one becomes aware of one's own self. In this theory of recognition, identity is not solitary and does not come from within the person. The theory of recognition emphasizes that identity formation occurs in intersubjective relations and hence is based on the mutual recognition between individuals.

In the work of Honneth the concept of recognition and the ideas of identity formation are developed further by distinguishing three separate spheres of social interaction: the private, the legal and the communal. These three spheres relate to three kinds of identity formation or self-relation, for which Honneth uses the following terms:

- Private (love, family, friendship): self-confidence
- Legal (rights): self-respect
- Community (shared value orientation): self-esteem

Self-confidence derives from affective relations, self-respect is gained in juridical relations and self-esteem is maintained from shared value orientations. These three terms are employed by Honneth for very distinct purposes and are thus "artificial" or "technical" terms in the sense that they are used to describe and analyse three different forms of self-understanding related to the three spheres of recognition. The terminology is "artificial", as in everyday language these terms are not used in this distinctive manner.

Recognition, to begin with, is the recognition of the autonomy of the others. Therefore, misrecognition is the opposite. Misrecognition is expressed by the showing of contempt, disrespect, disregard or even disdain for another person. Disrespecting the autonomy of another person would imply a violation of that person in one or more of the three layers of autonomy proposed by Hill and mentioned in Section 3: moral philosophy, legal rights and theory of identity. Disrespecting autonomy implies considering a person to be incapable of conducting independent moral judgements; it would be a violation of a person's legal right to independence and a rejection of the specific identity of a person as valuable and legitimate. The injury or violation plays a crucial role in Honneth's philosophy of recognition, as is also the case in Hegel, because it is through the violation of the legal forms of recognition that the intersubjective nature of human relations comes into consciousness.[323] For this reason, in order to understand recognition it is useful to start by understanding what is misrecognition.

[323] Honneth, Axel: *Kampf um Anerkennung*, p. 43.

The concept of *consciousness of injustice* (German: *Unrechtsbewusstheit*) is used by Honneth to describe how ethics work among the "lower strata", as he calls them. This ethic of ordinary people is not developed into a consistent, articulated and reflected system, as it is in the "qualified strata", but consists mostly of situation-related statements dominated by "negative morality", i.e. by the unjust.[324] The focus is thus not on the just but on the unjust. The unjust comes so to speak before the idea of the just. I, however, believe this goes for most people, academic or not. To understand what is just cannot be done without an understanding of the unjust. Honneth writes:

> Recognition expresses the fact that the other person is supposed to possess social 'validity'. . . . A decentering takes place in the recognizing subject because she concedes to another subject a 'worth' that is the source of legitimate claims infringing upon her own self-love.[325]

This understanding of ethical consciousness is not new; it can be found in the works of several philosophers beginning with Plato – who emphasized that we cannot understand the good except on the background of evil – and continuing up to the present. And it seems quite evident that good and evil are related. How could one have an idea of the good without an experience or idea of the evil? The theory is thus that positive knowledge of what is right and good presupposes knowledge of, or a certain sensibility towards, vices and evils. This is brought about by the experience of humiliation and injustice. Without the experience of so-called "negative morality" like suffering, envy, corruption, violation, cruelty, betrayal and so forth it would be impossible to think about and formulate positive values like autonomy, integrity, honesty, happiness, freedom, human rights, care for others and so on. It is also true that in general people are much more sensitive to injustice than justice. Jonathan Allen formulates it this way:

> We do not know what a perfectly just society would be like to live in, but we do know – some of us know from everyday experience – what it is to live in an unjust, or cruel, or humiliating society. . . it is often easier to identify evils than it is to recognize and understand goods.[326]

The articulation of positively stated norms has to take an indirect route, i.e. it must follow the route of "negative morality", which is expressed as vices, evils, injustices and so forth. The idea presented by Allen is that where evils are clearly evident, their removal or minimization has priority over the pursuit of goods, and that negative morality has an important role in formulating judgements.

[324] Honneth's development of the concept of *consciousness of injustice* is based on the works of Barrington More and E.P. Thomson and others. Honneth, Axel: "Moral consciousness and class domination: Some problems in the analysis of hidden morality". In Honneth, Axel: *The Fragmented World of the Social*, p. 209.

[325] Honneth, Axel: "Invisibility: On the Epistemology of 'Recognition'", p. 121.

[326] Allen, Jonathan: *The Place of Negative Morality in Political Theory*, p. 350.

An excellent illustration of how we should understand recognition by means of the concept of misrecognition can be found in Honneth's "Invisibility: On the Epistemology of 'Recognition'". Honneth opens by referring to the novel *Invisible Man* (1952) by Ralph Ellison. The narrator, a young, nameless black man, moves through American intolerance in the 1950s. "I am an invisible man. When they approach me they see only my surroundings, themselves, or figments of their imagination – indeed, everything and anything except me," he says in the prologue of the book. What the narrator is referring to here is the white ignorance of the black and how it was used as an intentional sign of lack of recognition. Black people were not regarded as autonomous individuals, they were considered inferior and subject to the ruling white regime. In the public sphere the actual physical presence of the blacks was ignored by the whites: the blacks were *invisible*. The whites looked through the blacks as if they were non-existent. The narrator in *Invisible Man* describes this as a humiliation or violation.

Honneth uses this in his metaphorical description of what it means to be recognized, because becoming visible – in a non-visual sense – is an elementary form of recognition. Recognition is dependent on expressive gestures from others indicating that one is "visible" and therefore socially validated. Without such gestures a person becomes invisible and thus excluded from the social sphere.

From this brief description of the theory of recognition according to Honneth I now turn to the two questions raised earlier to see how this applies to the market and ethical consumption. (1) What kind of recognition is involved in the consumers' recognition of the production history and the producers and retailers? (2) How can the individual autonomy of ethical consumers be recognized?

Starting where we ended our theoretical investigation on recognition, namely in the analysis of misrecognition, we can draw the analogy to modern industrial food production and distribution, which here and elsewhere has been characterized by its lack of transparency and traceability. Producers and production histories are invisible to the consumers – though we should note that the opposite is also true: consumers are invisible to the producers too. This lack of mutual visibility renders recognition on the food market at best difficult and in most cases impossible. Consumers cannot recognize producers and production histories of the food, and producers are unable to recognize consumers' claims concerning the production history.

Let us deal with the first question. In spite of the non-transparency of the market, some of the retailers, those who run shops or supermarket chains, actually *are* visible to the consumers. This is not just a trivial observation: when this visibility is so clearly expressed in the empirical material from the survey mentioned above, it is because the retailers are the only representatives of the market, and thus of the production history, that the consumers enter into contact with. Of course, this visibility is used by consumers to judge, evaluate, select and ignore certain shops according to personal preferences. In the case of consumption we cannot uncritically make the analogy from racist ignorance outlined above as a sign of misrecognition, because the kind of recognition in the two cases is quite different. Becoming visible in a "non-visual" sense does not mean that the autonomy of the producers is recognized by the consumers. The ignorance of the blacks was due to the whites' unwillingness to recognize their individual autonomy, and thus they were excluded from being fully valuable members of society. This is far from the case with ethical consumption. Choosing specific shops or foodstuffs can be seen as an expression of ignorance of other "non-chosen" shops and foodstuffs, but not as an expression of misrecognition of the autonomy of the shops or the producers.

What then, we need to ask, is recognized from the perspective of ethical consumption? My answer is that it is how the production history corresponds to the consumers' own values. The ethical consumers' recognition, or the opposite – ignorance – is primarily a measurement of how the production history matches consumer expectations concerning production practices. The ethical consumers' recognition belongs first and foremost in the third sphere of recognition, i.e. the shared value orientation. Ethical consumers evaluate a history, which in turn is an evaluation of the people involved in that history. The recognition of the persons involved in the production history is in general an evaluation of their values concerning food production practices.

If we pursue Honneth's theory of recognition, it is tempting to conclude that consumers' recognition of producers and retailers is of importance for their own self-esteem, and their feeling of social validity. Such recognition from the consumers would be desirable also from the producers' perspective, not only in economic terms but also in individual and social terms because of the link to self-esteem. Positive recognition of the producers and retailers, when the producers and retailers become visible, is a validation of their work, and thus of importance for their self-worth and pride in their work.

The second question is: How does the philosophy of autonomy and recognition apply to ethical consumers and consumption? Staying with Honneth's idea that invisibility equals lack of recognition, the moral injury to ethical consumers is that producers look through them: *ethical consumers are invisible*. In the three spheres of recognition, the private, legal and communal, it is not the private sphere of love and friendship that is of interest here. In the private sphere food can be a means of showing recognition or even disrespect, but on the market it is the legal and communal spheres that are relevant. The injury is not to the physical integrity of ethical consumers but to their legal procedural rights and to their social value (see below).

Ethical consumption may very well be an important element in some consumers' identity, but in the impenetrability of the market this identity is disregarded. Without positive recognition from others this identity in the legal sphere is emptied of self-respect. Lack of respect for the autonomy of the individual deters consumers from considering themselves trustworthy and responsible people. Their sensitivity on ethical issues is generally undervalued – probably because the market is believed to be a private sphere.

Furthermore, the lack of rights for ethical consumers makes the claim for ethical consumption illegitimate. When the demand for information about production history is not respected from a juridical point of view, it is very difficult to demand "the right to know" about it. This shows how important a role politicians and food authorities play in this respect. It is not too much to say that ethical consumption will remain in its niche without the support of politicians.

In the social sphere recognition is linked to "social value" (German: *soziale Wertschätzung*), which refers to the specific characters and actions of individuals and how they are appreciated by other members of the society. Individuals are considered socially valuable when their character or actions are conceived as "valuable" by the other members of society.[327] From this recognition comes the experience in the individual of being valuable for society; it confers self-esteem. But again, due to the lack of recognition of ethical consumption, we must state that the possibility of acquiring self-esteem from ethical consumption is at present almost illusionary. Those who would wish to consider themselves ethical and responsible consumers (and I would add, citizens) will soon realize that they are deluding themselves. Most consumers are therefore tempted to give up this identity and stick to the role of the ordinary consumer in the classic liberal sense of the word.

This may sound unlikely, since it is often stated that "the consumer is king" and there is no place in business for those who choose to ignore the consumers. In which sense then are ethical consumers ignored and disrespected? They are ignored because they are *seen* solely as utility-maximizing individuals. To recognize them as responsible and socially valuable would mean to provide them with information on the production history. And I mean *provide*, because ordinary consumers are not qualified to find this information, even in the rare cases when they demand it. It is therefore the responsibility of producers, retailers, politicians and food authorities to make the information accessible. Denying access to information on the production history, and not actively making it accessible, means to misrecognize consumers' capacity to take an ethical stance.

Knowledge of this history is a condition for the ethical judgement of the consumers. The etymological origin of the Greek verb *histereo* means (1) to learn by inquiry and (2) to narrate what one has learned. The noun *histor* is

[327] Honneth, Axel: *Kampf um Anerkennung*, p. 209.

found in Homer, and the Homeric "historian" is the judge.[328] The historical perspective is essential for the consumers in order to be able to *tell how it was* and to judge food from an ethical perspective. Writing on the word's origin, Arendt speaks of the "clash between the spectator and the actor. The spectacle before the spectator – enacted, as it were, for his judgement – is history as a whole".[329] If the spectacle before the consumers is the production history, making a judgement on the history makes them a part of that history; it puts, so to speak, the consumers into the production history, and it makes of them partial and political actors.

The exclusion of consumption from political participation in democracy hinders consumers from becoming part of "active citizenry". The current structure of the market, which to a large extent is decided upon by policymakers, is at present discouraging consumers and citizens from political participation. The dominating liberal approach to the market is that of limited intervention. The classical and widespread liberal procedure is to leave all manner of political and ethical issues to the unregulated market; these include animal welfare, fair trade and ecological sustainability. If articles produced with attention to these issues cannot be sold, it is assumed that it is because there is no support among the general public for it. What is forgotten here is that the market favours an egoistic, utility-maximizing attitude (why should I pay more if my neighbour does not?). It is also forgotten that information on the products, especially the difference *between* the products, is not provided voluntarily by the producers themselves. This makes it impossible for the consumers to see the differences between the products. A simple policy in favour of a liberal market enhances self-interest, competition and struggle. At the same time it decreases solidarity and ethical concern. Admittedly, it can be a good idea to leave certain things to the market; indeed there is even a positive effect from a participatory point of view in many citizens going to the market frequently as part of their daily life. But in my view this should never be done without them simultaneously having access to reasonable and impartial information allowing them to make informed and autonomous choices.

The autonomy of consumers presupposes recognition in all three layers of individual autonomy described by Hill in Section 3: moral philosophy, jurisprudence and the philosophy of the person. Protection of the individual autonomy of consumers, in the sense of securing their independence, belongs to the legal sphere. From the perspective of moral philosophy and the conception of the subject, the power of the idea of autonomy lies in its ability to address the importance of protecting the unstrained, reflexive judgements and actions of

[328] Arendt, Hannah: *Lectures on Kant's Political Philosophy*, p. 5. See also Chapter Two on the original use of the word "history".
[329] Ibid. p. 58.

citizens and consumers. The development of individual autonomy in the form of informed choice in the realm of consumption is central to the development of new forms of participation in democratic societies. The respect of consumers' autonomy functions as a strong motivational factor for the participation of consumers and citizens in societal matters, not necessarily in traditional politics but in the general social life. This is so because respecting individuals' autonomy makes it possible for them to make a difference through active participation. Thus, the idea of an autonomous person, i.e. a thinking, willing and acting person, first, contributes to the individual's self-esteem and self-understanding and, second, motivates social engagement. Thus, when considering consumption, respect of the consumers' autonomy could contribute to the consumers' self-esteem and self-understanding; consumers would not only be agents on the market but also have the possibility of becoming ethical and political agents. Also, because autonomy makes one the author of one's own actions, it also links up to responsibility.

Ethical consumption is based on the wish to contribute to the common good through consumption. It is also a vision of political participation of consumers and as such it relates to some extent to the radical conception of liberal democracy of the American philosopher John Dewey (1859–1952), e.g. in his book *The Public and Its Problems* (1927). Although Dewey did not consider consumption as a sphere for political participation, and related participation only to citizens rather than consumers, his ideas are nonetheless illustrative of what could be the consequences of recognizing ethical consumption as a legitimate political sphere. His writings are a defence of participatory or deliberative democratic ideals. His idea of democracy as a reflexive form of societal cooperation states that collective problem-solving provides a paradigm for political opinion formation. Successful problem-solving builds on the collaboration of actors who engage in deliberation and have free access to information. Participation in democracy, according to Dewey, is part of our individual freedom, yet at the same time this participation constitutes a link or relationship to others. Participation relates people in the community to each other. For Dewey democratic participation is not restricted to the political sphere and to representative democracy but applies to a wide range of social spheres, such as the workplace. Dewey believed that public discussion would reveal common interests, as discussion is educative and "forces a recognition that there are common interests".[330] In our concrete case of ethical consumption, it is reasonable to follow the line of the argument that consumers, retailers and producers, all have a common interest in, for instance, animal welfare, yet there is no common public sphere in which such an interest can be expressed and shared. The actors in the market are separated by market non-transparency. Retelling the production

[330] Dewey, John: *The Public and Its Problems*, p. 207.

history in public would generate an easily accessible sphere, where actors in the market can participate and present their opinions. The advantage of this setting is that it links up closely with the everyday life of every citizen, for we all have to eat. The political participation would then not be distanced from the matters in question – which I believe is often the case in representative democracy or in the work of political parties. The matters in question would be close to the citizen consumers, involved as they are through eating. The consumers' own opinions and judgements would easily be transformed into action through shopping behaviour. It would allow for exchange of opinions and even for disagreements, conflicts and differences in attitude, as it would probably leave space for foodstuffs with different production histories. Making the production history accessible to the public is therefore absolutely consistent with the idea of pluralism or with Dewey's pragmatism.

Had he dealt with consumption, I would guess that Dewey would have emphasized the educative aspect of consumers participating in the evaluation and judgement of the production history. This would imply moral development, self-development (or transformation of the self) in the sense described in the previous chapters on the interpretation of the production history. Dewey was well aware that equal participation in general requires that disparities in wealth are not too large.[331] It has often been stated that the idea of "voting with money" excludes the poor from political participation and favours the rich. Poor people cannot afford impartiality when judging a production history. We saw, however, that impartiality is important for the moral and autonomous judgement and therefore also for the choice. Poor people tend to be partial in the sense that they first of all must see to it that they do not go to bed hungry. It can be argued that as they can always only afford the cheapest food they have no real choice. This is a serious point and I would not argue against the importance of a reasonably equal access to wealth or just distribution. However, I do not think that it is an argument that should be used to repress ethical consumption. Considering that food prices have pro rata been decreasing constantly in many Western countries since the 1950s, it could also be argued that spending on food has reached a level where large segments of the population would actually be able to spend more if they had the motivation to do so. The fact that the purchase in Denmark of organic food, which is more expensive than conventional food, is largest among students underlines this; students clearly belong in the low-income class of society.

In conclusion, recognizing ethical consumers means to recognize their procedural claim for information, not only about ingredients but also about the ethical aspects of food production practices. This enables the consumers to recognize – or not to recognize – the *production history and those involved in it.* Producers can recognize the consumers' claim to information on the production

[331] Caspary, William R.: *Dewey on Democracy*, p. 12.

history – and by informed choice the consumers are able to recognize the producers. Or as the title of this section says, information on the production history may facilitate recognizing consumers recognizing producers.

The consumers then become "visible" in a non-visual sense, i.e. they are recognized by the producers and retailers through the communication of the production history. Producers also become "visible" to the consumers by telling them the production history. In both cases – producers recognizing the consumers' ethical identity and ethical consumers recognizing the producers' social validity and production practices – the production history or the transparency of the production process is the means to assuring the structural frames for a possible recognition.

However, cognition does not automatically lead to recognition. The recognition of the producers and retailers by the consumers is not assured automatically by making the producers *cognitive* or visible, i.e. by giving consumers access to the production history or traceability of the product. This negative reaction may come about in two ways: first, if consumers disapprove of the cognitive production history; and second, if consumers choose to ignore the production history and thus the social validity of the producers and food production practices from an ethical perspective.

The actual or current deregulation and market orientation of society has paved the way for new initiatives on information to consumers to be initiated by the business community. There are already telling examples of how the commitment of politicians and authorities has had a significant influence on the successful acceptance and recognition by consumers of the producers' situation and practices, e.g. in the labelling of organic products. So long as a political interest can be sustained in guiding society towards an acceptance of ethical opinions and the regard for people, society, nature and the environment, it is political action that is the crucial instrument.

If citizens feel that the representative political system is inadequate, and that it is difficult to make a difference, other methods can be put to use. There is a sense in which consumption has become a social experiment within democracy, a new way of expressing opinion, and there is reason enough to challenge existing democratic institutions and participatory forms. As such, consumption has both advantages and disadvantages in the pursuit of democracy.

A widespread criticism of the political consumer society, where cash power equals voting power, is that the rich acquire the greatest influence, while the poor are deprived of democratic influence. This is a powerful but hasty conclusion, for we must consider what is the ultimate purpose of ethical consumption. It is perhaps asking too much of it to claim it as a substitute for politics and democracy, but as a vitalization of democracy it is a powerful instrument inasmuch as we are buying things everyday of the week. And we must question whether it really is the case that the possibility of political and ethical consumption will make a greater difference between rich and poor than is the case at present. On the contrary, I would say, one could hope for the opposite. For the fact is that we do not know what would happen if ethical consumption became a genuine possibility for consumers. Political consumers, understood as those who wish to shop politically, do of course exist, but they are of no significance, since no social value is ascribed to them. The challenge from ethical and political consumers needs to be taken far more seriously. Consumers are from a political point of view of no value on their own, for someone has to make the necessary information available. Ethical consumers should be seen as a welcome challenge, because they force us to reflect on the role of the individual in a democracy, both as a citizen and as a consumer.

REFERENCES

Accum, Friedrich: *Culinary Chemistry*. Ackermann, London, 1821.

Allen, Jonathan: "The place of Negative Morality in Political Theory". *Political Theory*, 29(3), 2001, 337–363.

Allison, Henry: *Kant's Theory of Taste. A Reading of the Critique of Aesthetic Judgment*. Cambridge University Press, Cambridge, 2001.

Arendt, Hannah: "What is Existenz Philosophy?" *Partisan Review*, XIII(1), 1946, 34–56.

Arendt, Hannah: *On Violence*. In *Crisis of the Republic*. Harcourt Brace Jovanovich, New York, 1969.

Arendt, Hannah: *Lectures on Kant's Political Philosophy*. Edited by Ronald Beiner. The University of Chicago Press, Chicago, 1982.

Aristoteles: *The Complete Works of Aristotle*. Edited by Jonathan Barnes. Princeton University Press, Wat Sussex, 1984.

Bachelard, Gaston: *The Philosophy of No*. (1940). The Orion Press, New York, 1968.

Bachelard, Gaston: *The New Scientific Spirit*. (1934). Beacon Press, Boston, 1984.

Baranzke, Heike, Franz-Theo Gottwald and Hans Werner Ingensiep (eds.): *Leben, Essen, Töten. Anthropologische Dimensionen* (*Living, Eating, Killing*). Hirzel Verlag, Stuttgart, 2000.

Baudrillard, Jean: *The Consumer Society*. (1970). Sage Publications, London, 1998.

Beckett, J.V.: *The Agricultural Revolution*. Basil Blackwell Ltd, Oxford, 1990.

Benton, David: *Food for Thought*. Penguin Books, London, 1996.

Bergson, Henri: *Creative Evolution*, Holt and Company, New York, 1911. From French: *L'évolution créatrice*. (1907).

Böhme, Hartmut: "Transsubstantiation und symbolisches Mahl. Die Mysterien des Essens und Naturphilosophie". (*Transsubstantiation and the Symbolic Meal. The Mysteries of Eating and the Philosophy of Nature*). In: *Zum Naturbegriff der Gegenwart*. Kongreßdokumentation zum projekt *Natur im Kopf*. Stuttgart, 21–26 June 1993, hg. v. der Landeshauptstadt Stuttgart, Kulturamt, 1994, pp. 139–158.

Bonde, Niels, Jesper Hoffmeyer and Henrik Stangerup (eds.): *Naturens historiefortællere* (*Nature's Storytellers*). Gad, København, 1987.

Borges, Jorge Luis: *Other Inquisitions*. Square Press, New York, 1966. Original: *Otras inquisiciones*. Emecé, Buenos Aires, 1960.

Bourdieu, Pierre: *Distinction. A Social Critique of the Judgement of Taste*. (1979). Routledge, New York, 1986.

Brecht, Bertolt: *The Threepenny Opera*. (1928). Methuen Modern Plays, London, 1979.

Brillat-Savarin, Jean Anthelme: *The Physiology of Taste*. (1825). Counterpoint, Washington, 1949.

Brock, William H.: *Justus von Liebig. The Chemical Gatekeeper*. Cambridge University Press, Cambridge, 1997.

Canguilhem, Georges: *La connaissance de la vie*. (*Knowledge of Life*). Librairie Philosophique J. Vrin, Paris, 1952.

Canguilhem, Georges: *Etudes d'histoire et de philosophie des sciences concernant les vivant et la vie. (Essays on the History and Philosophy of the Life Sciences)*. Librairie Philosophique J. Vrin, Paris, 1968.

Caspary, William R.: *Dewey on Democracy*. Cornell University Press, Ithaca, 2000.

Castoriadis, Cornelius: *World in Fragments*. Edited and translated by David Ames Curtis. Stanford University Press, California, 1997.

Certeau, Michel de: *L'écriture de l'histoire*. (*The Writing of History*). Éditions Gallimard, Paris, 1975.

Chatwin, Bruce: *The Songlines*. Jonathan Cape, London, 1987.
Coff, Christian, Lise Walbom Christiansen and Eva Mikkelsen: *Forbrugere, etik og sporbarhed* (*Consumers, Ethics and Traceability*). Report from a survey conducted by Centre for Ethics and Law and *The Danish Consumers Co-operative Society* (FDB), 2005.
Crocker, David and Toby Linden (eds.): *Ethics of Consumption. The Good Life, Justice and Global Stewardship*. Rowman & Littelfield Publishers, Boston, 1998.
Curtis, Neal: *Against Autonomy. Lyotard, Judgement and Action*. Ashgate, Alderhot, 2001.
Darwin, Charles: *The Origin of Species*. First edition, Penguin Books, London, 1859. Reprinted 1985.
Descartes, René: *Discourse on Method*. (1637). Hackett Publishing Company, Indianapolis, 1993.
Descartes, René: *Meditations on First Philosophy*. (1641). Hackett Publishing Company, Indianapolis, 1993.
Dewey, John: *The Public and Its Problems*. (1927). Swallow Press, Athens, 1954.
Douglas, Mary: *Purity and Danger. An Analysis of Concepts of Pollution and Taboo*. Routledge, London, 1966.
Douglas, Mary: *Thought Styles*. Sage Publications, London, 1996.
Dyrberg, Torben Bech: *The Circular Structure of Power. Politics, Identity, Community*. Verso, London, 1997.
Elias, Norbert: *The Civilizing Process*. (1939). Blackwell Publishing, Oxford, 2000.
Fairclough, Norman: *Language and Power*. Longman, London, 1989.
Farb, Peter and George Armelages: *Consuming Passions. The Anthropology of Eating*. Houghton Mifflin Company, Boston, 1980.
Fernández-Armesto, Felipe: *Truth. A History and a Guide for the Perplexed*. Transworld Publishers, New York, 1997.
Ferry, Luc: *The New Ecological Order*. (1992). The University of Chicago Press, Chicago, 1995.
Feuerbach, Ludwig: *Die Naturwissenschaft und die Revolution*. (1850). Sämtliche Werke B.X., Kleinere Schriften III (1846–1850), Frommann, Stuttgart, 1982, pp. 347–368.
Foucault, Michel: *The Order of Things*. (1966). Routledge, London, 1970.
Foucault, Michel: *The History of Sexuality*. Vol. 1. (1976). Vintage Books, New York, 1978. Original title: *Histoire de la sexualité, I. La volonté de savoir*.
Foucault, Michel: *The History of Sexuality*. Vol. 2. (1984). Vintage Books, New York, 1985. Original title: *Histoire de la sexualité, II. L'usage des plaisirs*.
Freuchen, Peter: *Book of the Eskimo*. The World Publishing Company, Cleveland, Ohio, 1961, p. 154.
Gadamer, Hans-Georg: *Truth and Method*. Second edition. Sheed & Ward, London. 1975.
Giddens, Anthony: *The Consequences of Modernity*. Polity Press, Cambridge, 1990.
Goften, Leslie: "Bread to Biotechnology; Cultural Aspects of Food Ethics". In Mepham, B. (ed.): *Food Ethics*. Routledge, London, 1996, pp. 121–122.
Grigg, David: "Agricultural Trade and Agricultural Systems; Changes in the Types of Farming c. 1850–1930." In Havinden, M.A. and T. Collins (eds.): *Agriculture in the Industrial State*. Rural History Centre, University of Reading, 1995, pp. 44–59.
Grimod de la Reynière, Alexandre-Balthazar-Laurent: *Almanach des Gourmands. Huitième année (1812)*. (*Almanac of Gourmets*). Le petit Mercure, Paris, 2003.
Gronow, Jukka and Alan Warde (eds.): *Ordinary Consumption*. Routledge, London, 2001.
Heidegger, Martin: *Being and Time*. (1926). Harper and Row, New York, 1962.
Hill, Thomas E.: *Autonomy and Self-respect*. Cambridge University Press, Cambridge, 1991.
Hoffmeyer, Jesper: *Signs of Meaning in the Universe*. Indiana University Press, Bloomington, 1998.
Højrup, Thomas: *Det glemte folk*. (*The Forgotten People*). Institut for europæisk folkelivsforskning. Statens Byggeforskningsinstitut, København, 1983.
Holderness, B.A.: "Apropos the Third Agricultural Revolution. How Productive was British Agriculture in the Long Boom, 1954–1973?" In Mathias, Peter and John A. Davis (eds.): *Agriculture and Industrialization*. Blackwell Publishers, Oxford, 1996.
Holy Bible, New International Version. Hodder & Stoughton, London, 1984.
Honneth, Axel: *Kampf um Anerkennung. Zur moralischen Grammatik sozialer Konflikte*. (*The Struggle for Recognition. The Moral Grammar of Social Conflicts*). Suhrkamp Varlag, Frankfurt am Main, 1992.
Honneth, Axel: *The Fragmented World of the Social: Essays in Social and Political Philosophy by Axel Honneth*. Edited by Charles W. Wright. State University of New York Press, New York, 1995.
Honneth, Axel: Invisibility. "On the Epistemology of 'Recognition'". In *The Aristotelian Society*, Supplementary Vol. LXXV, Blackwell Publishers, Bristol, 2001, pp. 111–126.

Horkheimer, Max and Adorno, Theodor W.: *Dialektik der Aufklärung*. (*Dialectic of Enlightenment*). Fischer, Frankfurt am Main, 1947.
Horsdal, Marianne: *Livets fortællinger – en bog om livshistorier og identitet*. (*Stories of Life – a Book about Lifestories and Identity*). Borgen, København, 1999.
Isager, Sigen and Jens Erik Skydsgaard: *Ancient Greek Agriculture*. Routledge, London, 1992.
Jacob, François: *The Logic of Life. A History of Heredity*. Princeton Science Library, New Jersey, 1973. Translated from *La logique du vivant. Une histoire de l'hérédité*. Gallimard, Paris, 1970.
Jensen, Hans Jørgen Winther: *Diffusionen af ny teknologi på bondebrug i Danmark 1800–1915*. (*The Diffusion of New Technology in Agriculture in Denmark 1800–1915*). Ph.D. dissertation, Copenhagen, 1998.
Johanssen, Wilhelm: *Elemente der exakten Erblichkeitslehre*. (*Elements of Heredity*). København, 1909.
Jonas, Hans: *The Phenomenon of Life. Toward a Philosophical Biology*. (1963). Delta, New York, 1966.
Jonas, Hans: *The Imperative of Responsibility*. (1979). University of Chicago Press, Chicago, 1994. Translation of *Das Prinzip Verantwortung*. Suhrkamp, Frankfurt am Main, 1979.
Jonas, Hans: *Das Prinzip Leben*. (*The Principle of Life*). Suhrkamp, Frankfurt am Main, 1997.
Jørgensen, Sven Erik: *Integration of Ecosystem Theories: A Pattern*. Kluwer Academic Publishers, Dordrecht, 1992. Revised edition 1997.
Kant, Immanuel: *The Critique of Judgement*. (1790). Oxford University Press, Oxford, 1952.
Kant: *First Introduction to The Critique of Judgment*. (1789/1790). The Library of Liberal Arts, Indianapolis, 1965.
Karafyllis, Nicole C.: "Renewable Resources and the Idea of Nature – What Has Biotechnology Got to Do With It?" *Journal of Agricultural and Environmental Ethics*, 16, 2003, 3–28.
Kay, James and E.D. Schneider: "Thermodynamics and measures of ecological integrity". In: *Proceedings of Ecological Indicators*. Elsevier, Amsterdam, 1992, pp. 159–182.
Kemp, Peter: *L'irremplaçable. Une Éthique de la technologie*. (*The Irreplaceable. An Ethics of Technology*). Les Éditions du Cerf, Paris, 1997.
Kemp, Peter: *Tid og fortælling. Introduktion til Paul Ricœur*. (*Time and Narrative. An Introduction to Paul Ricœur*). Aarhus Universitetsforlag, Århus, 3. udgave, 1999.
Kemp, Peter: "Four Ethical Principles in Biolaw". In Kemp, Peter, Jacob Rendtorff and Niels Mattsson Johansen (eds.): *Bioethics and Biolaw*. Rhodos, Copenhagen, 2000, pp. 11–22.
Kemp, Peter: *Praktisk visdom. Om Paul Ricœurs etik*. (*Practical Wisdom. On the Ethics of Paul Ricœur*). Forum, København, 2001.
Kettlitz, Beate: "Consumer perception of food risk and consumer trust". EurSafe's Third Congress on *Food Safety, Food Quality and Food Ethics*, 2001, pp. 51–59.
Kjærgaard, Thorkild: *Den danske revolution 1500–1800. (The Danish Revolution 1500–1800)*. Gyldendal, København, 1991.
Korthals, Michiel: "Taking Consumers Seriously. Two Concepts of Consumer Sovereignty". *Journal of Agricultural and Environmental Ethics*, Kluwer Academic Publishers, 14, 2001, 201–215.
Korthals, Michiel: *Before Dinner. Philosophy and Ethics of Food*. Springer, Dordrecht, 2004.
Koselleck, Reinhart: *Futures Past. On the Semantics of Historical Time*. (1979). MIT Press, Cambridge, 1985.
Lavoisier, Antoine Laurent: *Traité de chimie*. Paris, 1793.
Lees, Michèle (ed.): *Food authenticity and traceability*. Woodhead Publishing Limited, Cambridge, 2003.
Lévinas, Emmanuel: *Totality and Infinity*. (1961). Martinus Nijhoff Publishers, The Hague, 1979.
Lévinas, Emmanuel: "La trace de l'Autre". (The Trace of the Other). In *En découvrant l'existence avec Husserl et Heidegger*. Vrin, Paris, 1963.
Lakoff, George: *Women, Fire and Dangerous Things. What Categories Reveal about the Mind*. The University of Chicago Press, Chicago, 1987.
Linné, Carl von: *Systema naturae*. Leyden, 1756, p. 215.
Luhmann, Niklas: *Trust. A Mechanism for the Reduction of Social Complexity*. (1968). John Wiley & Sons, Chichester, 1979.
Lukes, Steven: *Power: A Radical View*. Macmillan Press, London, 1974.
Lupton, Deborah: *Food, the Body and the Self*. Sage Publications, London, 1996.
MacIntyre, Alasdair: *After Virtue*. Second edition. Duckworth, London, 1981.
Malthus, Thomas Robert: *An Essay on the Principle of Population*. (1798). Cambridge University Press, New York, 1989.
Martin, Luther H., Huck Gutman and Patrick H Hutton (eds.): *Technologies of the Self. A Seminar with Michel Foucault*. University of Massachusetts Press, Cambridge, Massachusetts, 1988.

Marx, Karl: *Capital*. Vol. One. 1887.
Mellemgaard, Signe: *Kroppens natur. Sundhedsoplysning og naturidealer i 250 år*. (*The Nature of the Body*). Museum Tusculanums Forlag, København, 1998.
Mepham, Ben (ed.): *Food Ethics*. Routledge, London, 1996.
Merleau-Ponty, Maurice: *Phenomenology of Perception*. (1945). Routledge, London, 1989.
Meyer, Rolf and Arnold Sauter: "Bedürfnis-und Konfliktfeld Ernährung: Handlungsfelder für die Politik". Büro für Technikfolgen-abschätzung beim Deutschen Bundestag. (TAB), Brief Nr. 20 June 2001.
Michaelsen, Kim Fleischer: "Probiotika – Bakterier, der fremmer sundheden". (Probiotics – Bacteria Promoting Health). *Naturens verden*, 2002/2003, pp. 37–42.
Misztal, Barbara A.: *Trust in Modern Societies*. Polity Press, Cambridge, 1996.
Mittelstraß, Jürgen (ed.) *Enzyklopädie Philosophie und Wissenschaftstheorie*. (*Encyclopaedia of Philosophy and Philosophy of Science*). Verlag J.B. Metzler, Stuttgart, 1996, pp. 1–5.
Moberg, Harald: *Jordbruksmekaniseringen i Sverige under tre sekel*. (*Mechanization of Swedish Agriculture over Three Centuries*). Kungl. Skogs-og Lantbriksakademien, Stockholm, 1989.
Mohr, Hans: *Natur und Moral. Ethik in der Biologie*. (*Nature and Morality. Ethics in Biology*). Wissenschaftliche Buchgesellschaft, Darmstadt, 1987.
Moore, George Edward: *Principia Ethica*, Cambridge University Press, Cambridge, 1903.
Nerheim, Hjørdis: *Estetisk rasjonalitet. En analyse av konstitusjonsbegrepet i Kants Kritik der Urtheilskraft*. (*Aesthetic Rationality. An Analysis of the Concept of "Constitution" in Kant's Critique of Judgement*). Solum Forlag, Oslo, 1991.
Nietzsche, Friedrich: *The Gay Science*. (1882–1887). Vintage Books, New York, 1974.
Nietzsche, Friedrich: *Twilight of the Idols, or, How to Philosophize with a Hammer*. (1889). Oxford University Press, Oxford, 1998.
Nietzsche, Friedrich: *On the Genealogy of Morality*. (1887). Hackett Publishing Company, Indianapolis, 1998.
Nordisk Ministerråd: *Safety Assessment of Novel Food Plants. Chemical Analytical Approaches to the Establishment of Substantial Equivalence*. Nordic Council of Ministries, TemaNord 1998, 591, Copenhagen, 1998.
Nordisk Ministerråd: *Forbrugernes fornemmelse for etik*. (*Consumers' Sense of Ethics*). TemaNord 2001, 583, København, 2001.
Nordic Council of Ministers: *Food Labelling: Nordic Consumers' Proposals for Improvements*. TemaNord 2001, 573, Copenhagen, 2001.
Ohnuki-Tierney, Emiko: "We eat Each Other's Food to Nourish our Body. The Global and the Local as Mutually Constituent Forces". In Grew, Raymond (ed.): *Food in Global History*. Westview Press, Colorado, 1999, pp. 240–272.
O'Neill, Onora: *Autonomy and Trust in Bioethics*. Cambridge University Press, Cambridge, 2002.
Onfray, Michel: *Le ventre des philosophes. Critique de la raison diététique*. (*The Stomach of Philosophers*). Grasset, Paris, 1989.
Onfray, Michel: *La raison gourmande. Philosophie du goût*. (*The Reasoning of Gourmands. The Philosophy of Taste*). Grasset, Paris, 1995.
Oswald, John: *The Cry of Nature or, An Appeal to Mercy and to Justice, on Behalf of the Persecuted Animals*. In Morton, Timothy (ed.): *Radical Food. The Culture and Politics of Eating and Drinking 1790–1820*. Vol. 1. Routledge, London, 2000, pp. 143–170.
Overton, Mark: *Agricultural Revolution in England*. Cambridge University Press, Cambridge, 1996.
Pedrot, Philippe (ed.): *Traçabilité et responsabilité*. (*Traceability and Responsibility*). Economica, Paris, 2003.
Proust, Marcel: *In Search of Lost Time. Swann's Way*. Vol. One. Translated by C.K. Scott Moncrieff and Terence Kilmartin. Revised by D.J. Enright. Modern Library, New York, 1992.
Raine, Kathleen: "William Blake – Prophet of Imagination". *Resurgence*, 204, 2001, 6–9.
Ricœur, Paul: *Time and Narrative*. Vols I–III. (1983–1985). The University of Chicago Press, Chicago, 1984–1988.
Ricœur, Paul: *From Text to Action. Essays in Hermeneutics, II*. Northwestern University Press, Illinois, 1991.
Ricœur, Paul: *Oneself as Another*. (1987). The University of Chicago Press, Chicago, 1992.
Rilke, Rainer Maria: *Worpswede. Monographie einer Landschaft und ihrer Maler*. (1902). Carl Schünemann/Verlag/Bremen, 1987.

Roche, Daniel: *Histoire des choses banales. Naissance de la consommation XVIIe–XIXe siècle*. (*The Story of Everyday Items. The Birth of Consumption 1600–1800*). Fayard, France, 1997.
Roos, Jørgen: *Recollections*. Net-Bog-Klubben, Copenhagen, 1998.
Rousseau, Jean-Jacques: *Emile or On Education*. (1762). Penguin Books, London, 1979.
Saint-Exupéry, Antoine de: *The Wisdom of the Sands*. (Posthume 1948). The University of Chicago Press, Chicago, 1979.
Sassatelli, Roberta: "Tamed Hedonism: Choice, Desires and Deviant Pleasures". In Gronow, Jukka and Alan Warde (eds.): *Ordinary Consumption*. Routledge, New York, 2001, p. 95.
Schattschneider, Elmer Eric: *The Semi-Sovereign People: A Realist's View of Democracy in America*. Holt, Rinehart & Winston, New York, 1960.
Schmidt, Alfred: *The Concept of Nature in Marx*. (1962). London, 1972.
Schmidt, Lars-Henrik: *Viljen til orden*. (*The Will to Order*). Modtryk, Århus, 1988.
Schmidt, Lars-Henrik: *Smagens analytik*. (*Analytics of Taste*). Modtryk, Århus, 1991.
Seel, Martin: *Eine Ästhetik der Natur*. (*Nature's Aesthethics*). Suhrkamp, Frankfurt am Main, 1991.
Simmel, Georg: "Sociology of the meal". (1910). In Frisby, David and Mike Featherstone (eds.): *Simmel on Culture. Selected Writings*. Sage, London, 1997, pp. 130–135.
Smith, L. John: *The Psychology of Food and Eating*. Palgrave, Hampshire, 2002.
Sterelny, Kim and Paul E. Griffiths: *Sex and Death. An Introduction to Philosophy of Biology*. The University of Chicago Press, Chicago, 1999.
Taylor, Charles: *The Ethics of Authenticity*. Harvard University Press, Cambridge, 1991.
Thompson, Michael: "The Development of the Agricultural Substructure". In Havinden, M.A. and T. Collins (eds.): *Agriculture in the Industrial State*. Rural History Centre, University of Reading, Reading, 1995, pp. 146–164.
Thompson, Paul B.: *Food Biotechnology in Ethical Perspective*. Blackie Academic & Professional, London, 1997.
Thoreau, Henry David: *Walking*. 1862.
Thope, I. J.: *The Origins of Agriculture in Europe*. Routledge, London, 1996.
Van der Ploeg, Jan Douwe: "The Reconstitution of Locality. Technology and Labour in Modern Agriculture". In Marsden, Terry, Lowe, P. and Wattmore, S. (eds.): *Labour and locality; uneven development and the rural labour process*. David Fulton, London, 1992, pp. 19–43.
Vieira, Ernest: *Elementary Food Science*. 4th edition, Chapman and Hall, New York, 1996.
Warde, Alan: *Consumption, Food and Taste. Culinary Antinomies and Commodity Culture*. Sage Publications, London, 1997.
WHO: *Strategies for Assessing the Safety of Foods Produced by Biotechnology*. Report of a Joint FAO/WHO Consultation, WHO, Geneva, 1991.
Zwart, Hub: "A Short History of Food Ethics". *Journal of Agricultural and Environmental Ethics*, 12, 2001, Kluwer Academic Publishers, pp. 113–126.

INDEX